Transistor Circuit
Techniques
discrete and integrated

TUTORIAL GUIDES IN ELECTRONIC ENGINEERING

Series editors
Professor G.G. Bloodworth, *University of York*
Professor A.P. Dorey, *University of Lancaster*
Professor J.K. Fidler, *University of York*

This series is aimed at first- and second-year undergraduate courses. Each text is complete in itself, although linked with others in the series. Where possible, the trend towards a 'systems' approach is acknowledged, but classical fundamental areas of study have not been excluded. Worked examples feature prominently and indicate, where appropriate, a number of approaches to the same problem.

A format providing marginal notes has been adopted to allow the authors to include ideas and material to support the main text. These notes include references to standard mainstream texts and commentary on the applicability of solution methods, aimed particularly at covering points normally found difficult. Graded problems are provided at the end of each chapter, with answers at the end of the book.

Transistor Circuit Techniques

discrete and integrated

Third edition

G. J. Ritchie
Department of Electronic Engineering
University of Essex,

CHAPMAN & HALL/CRC

A CRC Press Company
Boca Raton   London   New York   Washington, D.C.

Visit the CRC Press Web site at www.crcpress.com

© 1993 by G. J. Richie

No claim to original U.S. Government works
International Standard Book Number 0-7487-4075-9
Printed in the United States of America   3 4 5 6 7 8 9 0
Printed on acid-free paper

Contents

Preface to the third edition

Many encouraging comments extending back to the launch of the First Edition have prompted additional chapters on audio power amplifiers and power supplies. Naturally, new concepts are introduced but many of the techniques covered in earlier chapters are reinforced, particularly by the three substantial design studies.

Again, as for the Second Edition, the opportunity has been taken to rationalize and update references to other books, in particular to those in this series.

I gratefully acknowledge useful discussion on audio amplifiers with Dr Malcolm Hawksford, as well as the careful and constructive comment from my editor, Professor Greville Bloodworth.

G.J. Ritchie

Preface to the second edition

A penalty suffered by the author of the first book in a series is his inability to refer to those that follow. Now with a substantial number of books published in the series, it is possible in this second edition to cross-reference many of these excellent texts. Several new problems have been added and, by popular request, the material on *h*-parameters has been extended in the form of an additional Appendix. I am very grateful to Professor John Sparkes (author of the title 'Semiconductor Devices' in this series) for his detailed comments and revisions suggested to harmonize our efforts.

Preface to the first edition

It has been my experience in teaching electronic circuit design that many first-year degree students are frustrated by the lack of suitable texts at the right level of practical and theoretical content. Introductory volumes tend to be rather elementary while authoritative reference texts prove too extensive for this sensitive audience.

In this book my aim has been to guide the student gently through the analysis and design of transistor circuits, providing worked examples and design examples as illustration. Spread liberally throughout each chapter are exercises to test the reader's grasp of the material and a set of problems at the end of each chapter provides useful and realistic assessment. Extensive use has been made of margin comments to reinforce the main text by way of highlighting the most important features, giving references for further reading, recalling earlier material, summarizing the approach and emphasizing practical points.

It was considered essential to introduce, at an early stage, the concept of representing semiconductor devices by simple d.c. and a.c. models which prove so useful in circuit analysis. A brief description of semiconductors and device operation is justified in providing a basis for understanding diode and transistor behaviour, their characterization and limitations. Great importance is attached to a basic appreciation of integrated devices, bipolar and field-effect, particularly in terms of their matching and thermal tracking properties, as well as the fundamental economic law of integration, minimize chip area, which dictates the techniques used in modern circuit design.

A very simple model of the bipolar transistor is developed using a single resistor (r_{be}) and a current source (β_{h}). This is adequate for most low-frequency requirements; only when considering current sources has the r_{ce} parameter of the full hybrid-π equivalent circuit been invoked. The author does not favour the use of *h*-parameters since they are purely numbers and do not give the inherent prediction of parameter variation with bias current and current gain which is the forte of the hybrid-π and simple models.

A wide range of transistor circuitry, both linear and switching, is covered in terms of fundamental qualitative circuit operation followed by analysis and design procedure. No apology is made for the extensive analytic treatment of circuits presented in this text – practice in analysis and engendering familiarity with design procedures are essential facets of the training of an electronic circuit designer.

It is hoped that this book instils a sound foundation of concept and approach which, even in this most rapidly developing area of modern electronics, will prove to be of lasting value.

I am grateful to my colleagues at Essex University, in particular Professors G.B.B. Chaplin and J.A. Turner and Dr J.K. Fidler, for many useful discussions. I also wish to thank my Consultant Editor, Dr A.P. Dorey of Southampton University, for his enthusiasm and very constructive assistance with this project.

Introduction to semiconductor devices

1

Objectives

□ To define terms such as intrinsic (pure) and extrinsic (doped) semiconductors, majority and minority carriers.

□ To explain in simple terms how a semiconductor diode operates and how its d.c. characteristic is expressed analytically by the diode equation.

□ To approximate the d.c. behaviour of a forward biased diode to a constant voltage and represent its a.c. behaviour by the dynamic slope resistance.

□ To explain junction breakdown and how a breakdown diode can be used as a simple voltage stabilizer.

□ To describe the operation of a bipolar junction transistor (BJT).

□ To define the terms current gain, cut-off and saturation applied to a BJT.

□ To describe the structure of integrated circuit components – BJTs, resistors and capacitors.

□ To explain the value of the (planar) integrated circuit process in being able to produce components which are matched and whose parameters track with temperature.

In the design of electronic circuits it is important to know about discrete semiconductor devices such as diodes and transistors, their terminal properties and limitations. While device behaviour can be expressed in terms of complex equations, it is much more important to be able to characterize devices in the form of approximate, simple, a.c. and d.c. models which assist in both the analysis and design processes.

This chapter aims to develop a simple understanding of device operation and characterization which subsequently is applied to the design of amplifiers and switching circuits. Although the emphasis is on discrete components and fundamental circuit techniques, the influence of integrated circuit design is equally important.

The following general references are useful for this chapter:

Millman and Grabel (1987), Chapters 1–5.

Sparkes (1987).

Semiconductors

A pure or intrinsic semiconductor is conveniently recognized as having a conductivity between that of a metal and of an insulator although, as we shall see later, this is not the formal definition of the term. Many elements and compounds exhibit semiconductor properties but in this text we shall restrict our discussion to Group 4 elements such as silicon.

Fig. 1.1a shows a very simple representation of the covalent bonding between silicon atoms in a crystal lattice structure. At a temperature of absolute zero the valence electrons are very tightly bound into the structure; none are free for conduction and the resistivity of the material is very high, approaching that of a

GaAs, GaP and GaAlAs are particularly important as materials for optical devices such as light-emitting diodes, photodetectors and lasers. Germanium has largely been supplanted by silicon for diodes and transistors and is not used in integrated circuit fabrication.

A formal treatment of conduction mechanisms in semiconductors is beyond the scope of this text.

1

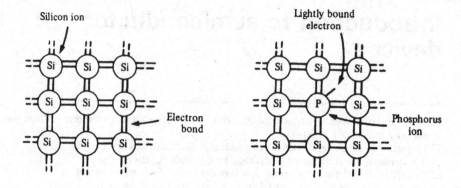

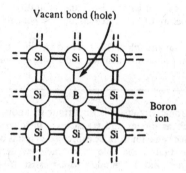

Fig. 1.1 (a) Pure silicon crystal (complete covalent bonding). (b) Phosphorus-doped n-type silicon (lightly bound electron). (c) Boron-doped p-type silicon (vacant bond = hole).

perfect insulator. However, as the temperature is raised the valence electrons gain more and more thermal (kinetic) energy and lose their immediate association with host ions; they become mobile and permit electrical conduction within the material. Thus resistivity falls with increasing temperature: a more correct definition of a semiconductor is – a material which exhibits a negative temperature coefficient of resistivity, at least over a certain temperature range. It is important to appreciate that the silicon ions are locked into the crystal lattice and, being immobile, do not contribute to the conduction mechanism.

In their pure crystalline state intrinsic semiconductors have little application to devices and are usually doped by the addition of a controlled amount of impurity.

If a Group 5 impurity element such as phosphorus is introduced, each phosphorus atom bonds covalently within the silicon crystal lattice and introduces one extra, lightly bound electron (Fig. 1.1b). These electrons take part in the conduction process at all but very low temperatures and are termed **majority carriers** in **n-type**, Group 5 doped semiconductors. The resistivity of a doped semiconductor is significantly less than that of the intrinsic material.

n-type → negatively charged electrons

In contrast, if a Group 3 element such as boron is introduced as impurity into the silicon crystal, the three bonding electrons of each boron atom form covalent bonds with adjacent silicon atoms leaving one vacant bonding site, or **hole** (Fig. 1.1c). A hole may be considered mobile, as an electron from a neighbouring atom can fill it leaving a vacant site behind; in this way, the hole has moved. It is convenient to think of holes as positively charged mobile carriers – majority carriers in Group 3 doped, **p-type** semiconductors.

p-type → positively charged holes

Doped semiconductors, both n-type and p-type, are also known as extrinsic semiconductors and the dopant ions, Group 3 or Group 5, are fixed in the crystal lattice just as are the silicon ions.

At normal ambient temperatures (around 290 K), mobile holes and electrons both exist in a semiconductor. However, the type of doping dictates which charge carrier dominates as the majority carrier (as described above), depressing below intrinsic level the concentration of the other carrier – the minority carrier. In n-type semiconductors, electrons are the majority carriers, holes the minority carriers; for p-type material, holes are the majority carriers and electrons the minority ones.

The junction diode

The simplest semiconductor component fabricated from both n-type and p-type material is the junction diode, a two-terminal device which, ideally, permits conduction with one polarity of applied voltage and completely blocks conduction when that voltage is reversed.

Consider a slice of semiconductor material one end of which is doped n-type, the other p-type. The n-type impurity dopant may be regarded as introducing fixed positively charged ions with loosely bound (negatively charged) electrons into the crystal lattice; the p-type dopant produces negative ions with attendant (positive) mobile holes.

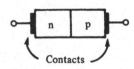

Contacts

Diode in equilibrium

In the immediate junction region between the n-type and p-type material, electrons can easily diffuse from the n-type into the p-type region, and holes can diffuse in the opposite direction. Both these diffusions result in a net transfer of positive charge from the p-region towards the n-region so that a potential difference and an electric field are developed between the two regions. The region within which this field is significant is called the **transition region**. In equilibrium the tendency of holes and electrons to diffuse and the effect of the field on the electrons and holes in the transition region just balance. The combined effect of both field and diffusion reduces the density of both electrons and holes to a level that is much less than the majority carrier density in either region, so the transition region is also sometimes called the **depletion layer**. It still contains small densities of mobile carriers so it is not wholly depleted; it also contains the positive and negative ions that are fixed in the lattice, as shown in Fig. 1.2, so transition region is usually the better term to use.

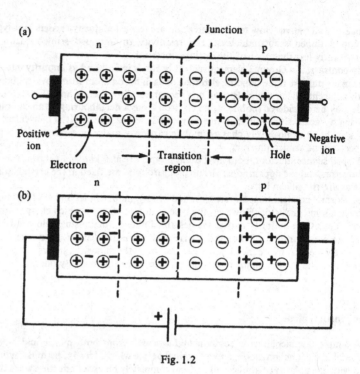

Fig. 1.2

Reverse bias

If an external potential is applied to the device, making the p-type material more negative with respect to the n-type, the electric field strength at the junction is increased, repelling mobile carriers further from the junction and widening the transition region (Fig. 1.2b). Under such circumstances, it would be expected that no current would flow across the junction with this reverse bias applied; however, in practice, a small current does flow. The **leakage** (or reverse) current is due to the **minority** carriers (the low-concentration holes in n-type and electrons in p-type) being attracted across the junction by the applied potential. It is temperature-dependent since, as the temperature is increased, more carriers are thermally generated.

In practice, because of surface leakage as well, it is reasonable to assume that the leakage current doubles approximately every 7 °C.

Variation of the width (w) of the transition region by applied voltage is important when considering the operation of junction field-effect transistors (see Chapter 7) and is given by

$$w \propto (\psi + V_r)^x \tag{1.1}$$

where V_r is the applied reverse voltage, ψ is the diffusion potential associated with the electric field at the junction ($\psi \approx 0.7$ V for silicon), and x is a constant, either 1/2 or 1/3 depending on the method used in fabricating the junction.

Forward bias

If the external bias voltage is now reversed, with the more positive potential connected to the p-region, the electric field strength in the transition region is reduced so that carriers can more easily flow through the junction. In a normal rectifier diode, holes from the p-region and electrons from the n-region flow through the junction. Since these opposing movements involve oppositely charged carriers they add together to form the total current I (amps) given by

$$I \propto \exp\left(\frac{qV}{kT}\right) \tag{1.2}$$

where q is the electronic charge (1.602×10^{-19} coulombs), V is the forward bias potential (volts), k is Boltzmann's constant (1.38×10^{-23} joule/K), and T is the temperature (K).

At a nominal ambient temperature of 290 K, kT/q can be evaluated as approximately 25 mV. This is an important figure, as will be seen later, and should be committed to memory.

The electron and hole currents (and the total current) may be regarded as the **injection** of majority carriers across the junction, the level of injection being controlled by the applied forward potential. The relative magnitudes of these current components are determined by the doping of the n-type and p-type regions. If the n-type region is much more heavily doped than the p-type then the forward current is almost all electron current; if the relative doping levels are reversed, the hole current is predominant. While this feature is of little significance with regard to the performance of junction diodes, it is vital in the manufacture of high-quality bipolar junction transistors.

The diode equation

The behaviour of a semiconductor junction diode may be summarized as

1. passing current under forward bias, with an associated forward voltage drop; and
2. exhibiting a very small leakage current under reverse bias.

This can be expressed as a **diode equation**:

$$I = I_s\left[\exp\left(\frac{qV}{kT}\right) - 1\right] \tag{1.3}$$

where I_s is the reverse leakage (or saturation) current. Fig. 1.3 shows this equation graphically (the device characteristic) and the diode symbol with defined directions of voltage and (positive) current.

Correspondence between the analytic expression of Equation 1.3 and the device characteristic can be checked. In the reverse region, for a sufficiently negative reverse voltage, the exp (qV/kT) term is very small and may be ignored relative to the (-1) term. Under this condition, the reverse leakage current is given by $I = I_s$. For a forward bias (V positive) of greater than 115 mV, the (-1) term has less than 1% significance and conveniently may be discarded leaving the forward bias region of the characteristic described by the approximate relationship

$$I = I_s \exp\left(\frac{qV}{kT}\right) \tag{1.4}$$

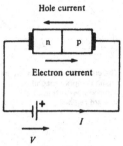

Hole current

Electron current

V

p-type positive and n-type negative for forward bias.

In correspondence with thermionic valve terminology, the p-type terminal is called the **anode** and the n-type the **cathode**.

Derivation of the diode equation is complex; the reader is asked to take it on trust or to consult specialized texts.

Equation 1.3 is a simplification of the full diode equation which contains, in the exponential term, an extra factor which is current and material-dependent.

The direction of current flow is conventionally defined as that of positive charge carriers despite the fact that the current may be electron current or, as here, the sum of electron and hole currents.

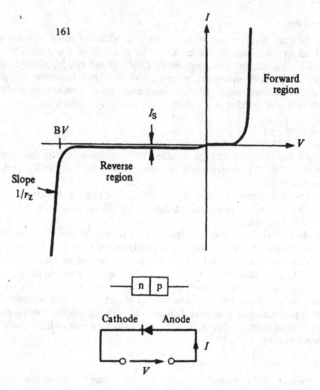

161

The current scale in the reverse region is highly magnified compared with that of the forward region.

Fig. 1.3 Junction diode characteristic (not to scale), symbol and defined current and voltage directions.

This corresponds with the injection description given by Equation 1.2.

The exponential nature of the forward characteristic makes it possible to calculate the change in forward voltage which results from increasing or decreasing the forward current by a certain ratio. Two useful ratios are the octave, a factor of 2 (or 1/2), and the decade, a factor of 10 (or 1/10).

There are corresponding voltages V_1 and V_2 for the two different currents I_1 and I_2.

It is interesting that we do not need to know the value of I_s to perform the voltage increment calculations. However, if we required the actual voltage, the value of I_s is necessary for calculation.

$$I_1 = I_s \exp\left(\frac{qV_1}{kT}\right)$$

and

$$I_2 = I_s \exp\left(\frac{qV_2}{kT}\right)$$

Therefore

$$\frac{I_2}{I_1} = \exp\left[\frac{q}{kT}(V_2 - V_1)\right] \tag{1.5}$$

6

or

$$(V_2 - V_1) = \frac{kT}{q} \ln\left(\frac{I_2}{I_1}\right) \tag{1.6}$$

If $I_2 = 2 \times I_1$, an octave relationship, then at $T = 290\,K$:

$$V_2 - V_1 = \frac{kT}{q} \ln 2 \approx 17.3\,mV$$

This implies that increasing the forward current by a factor of two increases the forward voltage by 17.3 mV irrespective of I_s and of the actual current level, provided that the (-1) term in the diode equation may be ignored. If $I_2 = 0.5I_1$, a halving of forward current, Equation 1.6 also shows that the forward diode voltage is reduced by 17.3 mV.

Note that the voltage increments are proportional to absolute temperature (K).

Now, for a decade change in current, $I_2 = 10 \times I_1$,

$$V_2 - V_1 = \frac{kT}{q} \ln 10 \approx 57.6\,mV \text{ at } 290\,K$$

and for a reduction in current by a factor of 10, i.e. $I_2 = 0.1I_1$,

$$V_2 - V_1 \approx -57.6\,mV$$

Another result of the sharply rising nature of the exponential forward characteristic, when it is plotted against linear current and voltage scales, is that there appears to be little conduction until (for a silicon diode) a voltage of approximately 0.5 V is reached. Above that voltage the current rises more and more rapidly such that, for normal operating currents, there is little change of forward voltage in the region of 0.7 V. This feature arises as a result of plotting the characteristic on linear scales; if diode voltage is plotted against the logarithm of the forward current, the characteristic becomes, over much of its length, a straight line with slope approximately 60 mV/decade.

Given that the forward voltage of a diode is 0.7 V for a forward current of 5 mA at a temperature of 290 K, calculate the reverse leakage current, I_s.

Exercise 1.1

[*Answer*: $I_s = 3.4 \times 10^{-15}$ A; a surprisingly low figure! In practice, low-power diodes usually exhibit leakage currents in the order of tens of nanoamps. The discrepancy between the two figures is due to current leakage across the physical surface of the diode which is additive to the junction leakage predicted by the diode equation. Another factor which destroys the exponential nature of the diode equation, particularly at higher current levels, is the resistance of the bulk doped semiconductor on either side of the junction; this gives an increased forward voltage at a given current.]

Temperature dependence of the diode characteristic can be determined by considering Equation 1.4 in the form

$$V = \frac{kT}{q} \ln \frac{I}{I_s} \tag{1.7}$$

Since we have already recognized that I_s increases with temperature then, to maintain a constant forward current I, the forward voltage V must be reduced as

This may seem to be an insignificant figure but it does represent 200 mV over a temperature range of 100 °C, a sizeable fraction of the normal forward voltage.

temperature is increased. Thus the forward voltage drop has a negative temperature coefficient which, in practice, is approximately $-2.2\,\text{mV}/°\text{C}$.

Breakdown

One feature of the diode characteristic not yet described is breakdown in the reverse bias region. When a certain reverse voltage, the reverse breakdown voltage (BV), is exceeded the reverse current increases dramatically for increasing reverse voltage (see Fig. 1.3) owing to the very high electric fields at the junction. The breakdown voltage can be controlled in manufacture by adjusting the doping levels: the higher the doping level, the lower the magnitude of the breakdown voltage. A diode with sufficiently high breakdown voltage should be chosen to preserve true rectifying action in normal circuit operation.

Depending on intended application, the breakdown voltage can range from several volts (breakdown diodes) to over 20 kV (high-voltage rectifiers).

Diode capacitance

While the nonlinear static (or d.c.) behaviour of a junction diode is characterized by the diode equation (Equation 1.3) or its approximation in the forward region (Equation 1.4) the device possesses capacitive properties which can be described in terms of transition capacitance and diffusion capacitance.

Transition capacitance: A junction diode under reverse bias may be considered as acting as a parallel-plate capacitor, the two plates being the bulk n-type and p-type semiconductor separated by the transition region dielectric. This transition capacitance (C_t) is proportional to the cross-sectional area (A) of the junction and inversely proportional to the width (w) of the transition region, i.e. the separation of the plates.

$$C_t \propto \frac{A}{w}$$

Since the transition width is a function of the applied reverse voltage as given by Equation 1.1, the transition capacitance is also a function of voltage

$$C_t \propto (\psi + V_r)^{-x}$$

which approximates to

$$C_t \propto V_r^{-x} \tag{1.8}$$

The capacitance of a varicap diode can be varied over a range of several hundred picofarads (large area device). Signal diodes generally have a capacitance of less than 10 pF.

(where $x = 1/2$ or $1/3$) for a reverse voltage (V_r) greater than several volts. Diodes used as voltage-variable capacitors (varicaps or varactors) find wide application in the tuning sections of radio and television receivers.

Diffusion capacitance: A junction diode also possesses capacitive properties under forward bias conditions by virtue of charge crossing the junction region. This is a complex concept and the reader should refer to more advanced texts for detail. However, it is sufficient to note that the diffusion capacitance (C_d) in forward bias is directly proportional to the forward current flowing through the device.

Diode ratings

Although semiconductor devices are robust and reliable, circuit designers must still ensure that they are operated within the range of capabilities for which they are manufactured. Diodes are no exception and information regarding maximum permissible parameter limits (or ratings) can be found published in manufacturers' data. The important factors for a diode are maximum reverse voltage (before breakdown), maximum forward current, and maximum power dissipation (the product of forward current and forward voltage). These ratings must not be exceeded otherwise device failure can result, with catastrophic consequences. Careful selection from a wide range of available device types is therefore essential for reliable design.

Imagine the consequences of failure in a nuclear power station, an aircraft navigation system or even a domestic television receiver!

Diode models

d.c. model

The diode equation with its exponential nature is very difficult to use directly in circuit analysis and design and it is useful to have an approximation to the characteristic which can provide a reasonably accurate indication of device behaviour.

In circuits using high voltages little error would result if a diode were assumed to be ideal, i.e. zero voltage drop in the forward direction and zero leakage current in the reverse direction. However the voltages in most semiconductor circuits are not very large and a forward biased diode voltage of approximately 0.7 V can prove very significant. Therefore, as a second level of approximation, it is realistic to assume a constant 0.7 V drop in the forward direction and again ignore leakage current in the reverse direction.

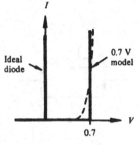

Exponentials and logarithms in equations are difficult to handle.

Small-signal a.c. model

When a diode is biased with a constant forward current (I) there is a corresponding voltage drop (V) across its terminals. If the current is changed by a small amount ($\pm \triangle I$) around I, the voltage will also change ($\pm \triangle V$) and for very small variations $\triangle I$ and $\triangle V$ are related by the tangential slope of the characteristic at the bias point (V, I). Owing to the curvature of the characteristic, this slope is not constant but varies with I; as I increases, the slope increases. It is useful to obtain an expression for this slope and its reciprocal (dV/dI) which has dimensions of resistance and is referred to as the dynamic slope resistance (r_d) of the diode.

Taking the approximate diode equation (Equation 1.4), and differentiating

This approach is used to simplify the subsequent mathematics which then becomes linear.

$$I = I_s\left[\exp\left(\frac{qV}{kT}\right)\right]$$

$$\therefore dI = \frac{q}{kT}I_s\left[\exp\left(\frac{qV}{kT}\right)\right]dV$$

$$= \frac{q}{kT} I\, dV$$

$$\therefore r_d = \frac{dV}{dI} = \frac{kT}{qI} \tag{1.9}$$

Now kT/q is approximately 25 mV at room temperature, hence the dynamic slope resistance can be expressed as

$$r_d \approx \frac{0.025}{I}\,\Omega \quad (I\text{ in A}) \tag{1.10}$$

9

or

$$r_d \approx \frac{25}{I} \, \Omega \quad (I \text{ in mA}) \tag{1.11}$$

This latter presentation (Equation 1.11) is the result which is normally used and clearly shows the dependence of slope resistance on the d.c. bias current (I).

Exercise 1.2 Calculate the dynamic slope resistance (r_d) of a diode, forward biased at the following currents: $10 \, \mu A$, $500 \, \mu A$, $1 \, mA$, $5 \, mA$.
[*Answer*: $2.5 \, k\Omega$, $50 \, \Omega$, $25 \, \Omega$ and $5 \, \Omega$ respectively.]

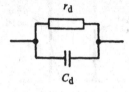

We are now able to represent the small-signal behaviour of a forward biased diode by its slope resistance (r_d) and, for high frequencies, include a parallel capacitance (C_d) representing its diffusion capacitance.

How small is a small signal? The trite answer is – vanishingly small, to preserve r_d as a constant slope over the signal excursion around the bias level. For other than zero amplitude signals, the slope of the characteristic changes, r_d, is not constant and the voltage/current relationship is nonlinear. It is customary, however, to use the model described above assuming a constant r_d but at the same time recognizing that nonlinearity (or distortion) increases with signal amplitude.

In Chapter 3 this small-signal representation of diode behaviour is developed to enable us to characterize the rather more complex bipolar junction transistor.

In the reverse bias region r_d is very high and may be omitted. We are then left with the diode being represented by its reverse bias transition capacitance (C_t) which degenerates, at low frequencies, to an open circuit.

Worked Example 1.1

The circuit of Fig. 1.4 shows a diode, a $10 \, k\Omega$ resistor and an a.c. voltage source connected in series across a $10.7 \, V$ d.c. supply. If the a.c. voltage source delivers a sinewave of $100 \, mV$ peak-to-peak amplitude, calculate the voltage across the diode.

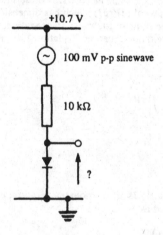

+10.7 V

100 mV p-p sinewave

10 kΩ

?

Fig. 1.4

Solution. The diode is in forward conduction since its arrow is in the direction of conventional current flow from positive to negative. Therefore, using the 0.7 V d.c. model, the average d.c. voltage across the diode is 0.7 V.

The d.c. voltage across the resistor is $(10.7 - 0.7)\,\text{V} = 10\,\text{V}$ and, since the resistor value is $10\,\text{k}\Omega$, the d.c. current through the diode is 1 mA.

The slope resistance of the diode is given by

$$r_\text{d} \approx \frac{25}{I}\,\Omega\,(I\ \text{in mA})$$

($25\,\Omega$ in this case).

This is an example of the application of the principle of superposition. The d.c. conditions (with a.c. sources turned down to zero) are evaluated first, then the a.c. behaviour is considered in isolation. The overall result is the additive superposition of the two cases since the a.c. signals are small and a linear model is used for the diode.

For the a.c. signal, the resistor and the slope resistance of the diode form a potential divider giving an a.c. diode voltage of

$$\frac{25\,\Omega}{25\,\Omega + 10\,\text{k}\Omega} \times 100\,\text{mV} \approx 250\,\mu\text{V peak-to-peak}$$

Therefore the diode voltage is an approximate sinewave of $250\,\mu\text{V}$ peak-to-peak amplitude superimposed on a d.c. level of approximately 0.7 V.

Breakdown diodes

Although reverse breakdown of a diode is a departure from its rectifying action, practical use can be made of this effect. If a diode is supplied with reverse current from a current source with a sufficiently high voltage capability ($> |BV|$), the diode voltage is substantially constant over a wide range of current. The diode, now used as a breakdown (or Zener) diode, has wide application in providing stabilized voltages ranging from 2.7 V to 200 V or more.

A breakdown diode is characterized by its nominal breakdown voltage and the reciprocal of the reverse characteristic in the reverse region, the dynamic slope resistance (r_z). An ideal breakdown diode has a well specified breakdown voltage and zero slope resistance giving a constant reverse voltage (in breakdown) independent of temperature and reverse current. In practice, however, the breakdown characteristic is curved in the low reverse current region and the reverse current supplied must be of sufficient magnitude to ensure that the breakdown diode operates beyond the knee of the characteristic in a region of low slope resistance. Further, even beyond the knee, slope resistance varies with reverse current and depends on the nominal breakdown voltage and temperature. Manufacturers' data should be consulted for accurate figures. In general, r_z is a minimum for devices with a $|BV|$ of approximately 6 V and operated at high reverse currents. At lower currents and for both higher and lower values of $|BV|$, r_z increases.

The temperature coefficient of breakdown voltage depends on both the nominal breakdown voltage and on the reverse current. Below approximately 5 V the temperature coefficient is negative and above is positive. This is because different breakdown mechanisms occur for low and high breakdown voltages. At approximately 5 V both mechanisms are present and produce a zero temperature coefficient.

The device rating which is important for a breakdown diode is the power dissipation, the product of reverse current and breakdown voltage.

Breakdown diodes are often referred to as **Zener** diodes and the breakdown voltage as the Zener voltage.

Voltage regulators are covered in detail in Chapter 9.

The nominal breakdown voltage is subject to a tolerance, e.g. 5.1 V, ± 5%.

Avalanche and Zener breakdown mechanisms are described in Chapter 2 of Sparkes.

Using a 400 mW breakdown diode and a resistor, design a simple stabilized voltage supply capable of providing 10 mA at 4.7 V from an existing + 12 V supply.

Design Example 1.1

...

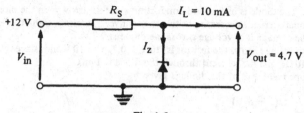

Fig. 1.5

Note the symbol for a breakdown diode is similar to that of a normal diode but with a tail on the cathode bar. In reverse conduction a breakdown diode is connected with the cathode positive.

Solution. The series resistor (R_S) limits the breakdown diode current which, although dependent on load current (I_L), allows the breakdown diode to develop a substantially constant output voltage (Fig. 1.5).

The output voltage is specified as 4.7 V; therefore a breakdown diode with a $|BV|$ of 4.7 V should be used.

Allow a minimum reverse current (I_z) of say 10 mA to flow through the breakdown diode, thus ensuring a reasonably low r_z.

The total current through R_S is $(I_L + I_z) = 20$ mA and the voltage across R_S is $(V_{in} - V_{out}) = (12 - 4.7) = 7.3$ V. Therefore

$$R_S = \frac{7.3}{0.02} = 365\,\Omega$$

It is not always appropriate to select the nearest preferred value. On occasion, the higher (or lower) adjacent preferred value may be the more suitable.

This is not a preferred value (see Appendix A), the nearest being 330 Ω and 390 Ω in the E12 series. The lower of these two values should be selected since the extra current reduces the slope resistance but it is essential to check that the power rating (400 mW) of the breakdown is not exceeded.

The power dissipation $= BV \times I_{z(max)}$. I_z is a maximum if the external load current were to fall to zero, i.e.

$$I_{z(max)} = \frac{V_{in} - V_{out}}{R_S} = \frac{12 - 4.7}{330} = 22\,\text{mA}$$

Therefore the dissipation is $(4.7 \times 0.022) = 104$ mW which is less than the rating of the breakdown diode. The design, with $R_S = 330\,\Omega$, is satisfactory.

Exercise 1.3

The superimposed a.c. behaviour of the circuit.

For the circuit of the preceding design example, if the 12 V supply is liable to variations of ± 0.5 V, calculate the voltage variation of the derived 4.7 V supply, given that the slope resistance of the breakdown diode is 40 Ω.
[*Answer*: ± 54 mV. This illustrates the ripple reduction of the simple voltage stabilizer.]

The bipolar junction transistor

A bipolar junction transistor (BJT) can be represented by a two-diode n-p-n or p-n-p structure as shown in Fig. 1.6, which also defines the symbols and the three terminals of the devices (emitter, base and collector) plus the terminal voltages and currents.

The direction of conventional current flow is that of positive charge carriers.

The arrow on the emitter lead serves two purposes. First, it distinguishes between the collector and emitter terminals which normally cannot be interchanged. Second, the arrow denotes the direction of conventional current flow through the device, pro-

12

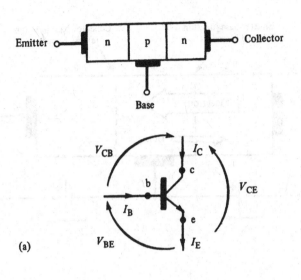

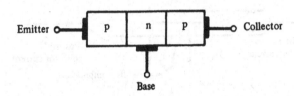

(a)

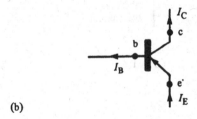

(b)

Fig. 1.6 Schematic structures, symbols, voltages and currents for (a) n–p–n and (b) p–n–p BJTs.

viding discrimination between the symbol for the n–p–n transistor and its p–n–p counterpart.

Figure 1.6 shows that for both n–p–n and p–n–p transistors,

$$I_E = I_C + I_B \qquad (1.12)$$

where I_E, I_C and I_B are the emitter, collector and base currents, respectively.

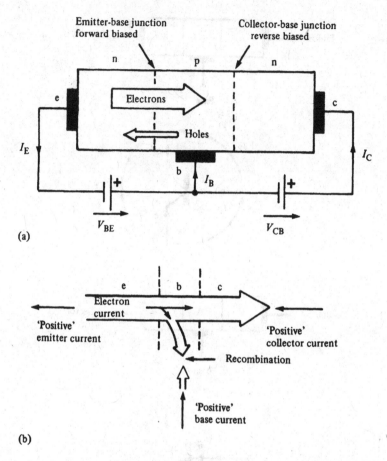

(a)

(b)

Fig. 1.7

BJT operation

In normal operation, the emitter-base junction is forward biased and the collector-base junction reverse biased. For the schematic n–p–n structure of Fig. 1.7a, electrons are injected from the n-type emitter into the base and, at the same time, holes are injected from base to emitter. To improve device efficiency, the doping level of the base region is made much lower than that of the emitter; essentially only electron current flows across the emitter-base junction with the injection level controlled (exponentially) by the base-emitter forward bias potential (V_{BE}), as

$$I_E \propto \exp\left(\frac{qV_{BE}}{kT}\right) \qquad (1.13)$$

14

Electrons injected from the emitter become minority carriers in the p-type base and, since the collector-base junction is reverse biased, these minority carriers which cross the base by diffusion are swept across the collector-base transition region. Because the electrons spend a finite time in transit through the base region, some recombine with holes; the holes involved in this recombination are replaced by positive charge flow into the base (via its connection to the bias source) resulting in a base current (I_B).

Transit time and recombination are reduced by making the base region very thin.

Operation of the transistor may be summarized by reference to Fig. 1.7b. The electron current at the collector is almost all of the current injected from the emitter diminished only by that lost as base current due to recombination. This consideration neglects hole injection from base to emitter and hole leakage from collector to base, both of which contribute to device current and hence degrade total efficiency.

BJT leakage current is discussed in the more expansive treatment by Millman and Grabel (1987), Chapter 3.

The term **bipolar** is applied to junction transistors of the type described above since two types of charge carrier (holes and electrons) are involved in the operation of the device. Unipolar, or field-effect, transistors (see Chapter 7) rely on only one type of carrier.

Why *bipolar*?

In this text leakage currents are ignored since, for silicon devices, they are significant only at elevated temperatures. (For consideration of leakage currents in devices and their effect on circuit design the reader should consult more advanced texts.)

BJT current gain

The fraction of the emitter current appearing as collector current is given the symbol α, the **common-base** current gain of the transistor – common-base since the base terminal is common to both the input port (emitter-base) and the output port (collector-base). Thus

$$\frac{\text{Collector current}}{\text{Emitter current}} = \frac{I_C}{I_E} = \alpha \tag{1.14}$$

The ratio (β) of the collector current to base current can be determined by combining Equations 1.12 and 1.14, as follows. Since

Many different symbols are used for various definitions of common-emitter current gain (β, β_F, β_o, h_{fe}, h_{FE} and so on). At this stage, it is unnecessary to distinguish between them.

$$I_E = I_C + I_B$$

$$I_B = (1 - \alpha)I_E$$

$$\frac{\text{Collector current}}{\text{Base current}} = \frac{I_C}{I_B} = \frac{\alpha}{1 - \alpha} = \beta \tag{1.15}$$

β is termed the common-emitter current gain, input at the base, output at the collector. The symbols h_{fb} and h_{fe} are widely used as alternatives to α and β (see Appendix C).

Rearrange Equation 1.15 to provide an expression for α in terms of β and hence determine α for a BJT whose measured β is 100.
[*Answer*: $\alpha = \beta/(1 + \beta) \approx 0.99$]

Exercise 1.4

For most bipolar transistors, α lies within the range 0.97 to 0.998, giving a corresponding β range of approximately 30 to 500. In other words, for a typical device, perhaps 1/100 of the emitter current is lost by recombination in the base. Since α is very close to unity it is often convenient to use the approximation

$$I_C \approx I_E \tag{1.16}$$

$$\approx I_s \exp \frac{qV_{BE}}{kT} \tag{1.17}$$

where I_s is a saturation or leakage current. I_C is nominally independent of the collector-base reverse bias (V_{CB}).

Although we have considered injection from emitter into base with the emitter as input, it is relevant to note, particularly when dealing with the common-emitter configuration, that the controlling input voltage is still the base-emitter voltage (V_{BE}) but in this case the input (base) current is only recombination current. If base current is supplied to the device, the base-emitter voltage and consequent emitter-to-base injection is forced to a level dictated by the base current and the current gain (β).

What about p–n–p transistors? This description of BJT operation has referred only to the n-p-n structure. A p-n-p device operates in exactly the same way except that holes are now the predominant charge carriers and the polarity of applied voltage and the directions of the terminal currents are reversed.

BJT characteristics

The input parameters for a BJT in common-emitter are the base-emitter voltage (V_{BE}) and the base current (I_B); the output parameters are the collector-emitter voltage (V_{CE}) and the collector current (I_C). The interrelation of these parameters can be presented graphically as device characteristics.

Current transfer characteristic: The current transfer characteristic of Fig. 1.8a, I_C plotted against I_B, is a graphical presentation of the common-emitter current gain (β). While this characteristic is relevant for a specific measured device, the very large variation (or spread) of β from one device to another, typically 10:1 even within a device type, limits general application. In representing current transfer by a straight line of slope β passing through the origin, two assumptions have been made; that leakage current is negligible and that β is a constant. Neither is true in practice; leakage currents can be significant at high temperatures and β falls off at both low and high currents, peaking to a maximum in the 1 to 10 mA range for low-power BJTs and at somewhat higher currents for high-power devices.

Input characteristic: A plot of I_B versus V_{BE} interrelates the two input parameters as the device input characteristic, shown in Fig. 1.8b. The exponential nature of the characteristic, expected from Equation 1.17 if β is assumed constant, is very similar to that of a diode and a constant voltage drop of 0.7 V is a satisfactory approximation to the V_{BE} of a conducting silicon transistor. This figure is widely used in the d.c. design of BJT circuits.

The same V_{BE} increments for a doubling of I_C and I_B (17.3 mV), and approximately 58 mV for a factor of ten times change in current, apply to the BJT. Also, the temperature coefficient of V_{BE} is approximately -2 mV/°C. All devices are not identical and V_{BE} is subject to a spread (typically ± 50 mV) for a particular device type.

Output characteristic: The graphical relationship between the output parameters, I_C and V_{CE} with I_B as control, provides the output characteristic shown in idealized form in Fig. 1.8c. To a first approximation it is valid to assume that the collector

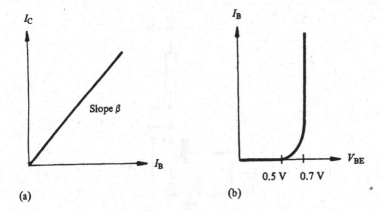

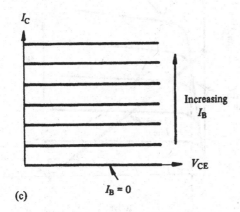

Fig. 1.8 n–p–n BJT characteristics: (a) current transfer, (b) input, (c) output (idealized).

current is determined only by I_B and β ($I_C = \beta \times I_B$) and is independent of V_{CE}; the BJT is therefore a current-controlled current source. However, the practical characteristic (Fig. 1.9) has non-zero slope implying that the collector is not a perfect current source (this is discussed further in Chapter 3). Also, curvature of the characteristic close to the I_C axis indicates that V_{CE} cannot fall to zero (except for zero current) as there must be at least a minimal residual voltage, termed the collector-emitter saturation voltage $V_{CE(sat)}$, which is typically 100 mV although it does depend on I_C, I_B and temperature.

It is fallacious to regard the BJT as a variable resistor controlled by I_B or V_{BE}. It is a controlled current source.

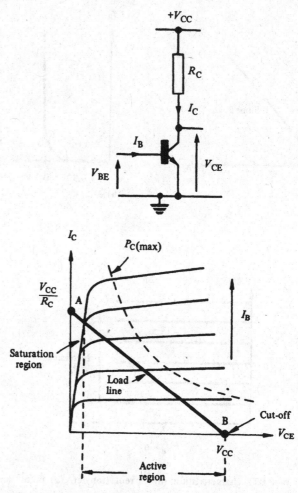

Fig. 1.9 Output characteristic with load line superimposed.

Load line

Consider a common-emitter BJT with the collector connected via a collector load resistor (R_C) to a positive voltage supply (V_{CC}) as shown in Fig. 1.9. Summing voltages gives

$$V_{CE} + I_C R_C = V_{CC} \tag{1.18}$$

or

$$I_C = \frac{V_{CC}}{R_C} - \frac{V_{CE}}{R_C} \tag{1.19}$$

18

Equation 1.19 can be superimposed on the output characteristic as a straight line of slope $-1/R_C$ intersecting the I_C axis at V_{CC}/R_C (point A) and the V_{CE} axis at V_{CC} (point B). This line is called the **load line** which, given V_{CC} and R_C, describes all possible operating conditions of the circuit.

Let us travel along this line and observe the conditions which apply. Starting at point B, for which $I_C = 0$ and $V_{CE} = V_{CC}$, the BJT is said to be cut-off or in the OFF state. This is achieved by open-circuiting the base terminal ($I_B = 0$) or by making $V_{BE} \leq 0\,V$ (for an n-p-n transistor).

If I_B (or V_{BE}) is raised slightly above zero, the BJT starts to conduct ($I_C > 0$) and enters the normal active region with the emitter-base junction forward biased and the collector-base junction reverse biased. Progressive increase of I_B (or V_{BE}) increases the collector current with a corresponding $I_C R_C$ reduction in V_{CE} (Equation 1.18) and the operating point moves along the load line towards point A.

As the collector voltage falls, the reverse bias on the collector-base junction reduces reaching zero bias when the collector and base voltages are equal. We are now at the other limit of the active region and about to enter the saturation region which is defined as the collector-base junction being forward biased. The approximately linear relationship between I_C and I_B no longer applies in this region and the curvature of the output characteristic limits I_C to a maximum of

The distinction is made here between saturation ($V_{CB} < 0$ for an n-p-n BJT) and the fully saturated or ON state.

$$I_{C(ON)} = \frac{V_{CC} - V_{CE(sat)}}{R_C} \tag{1.20}$$

where $V_{CE(sat)}$ has been described earlier. The BJT is now said to be fully saturated or ON.

BJT ratings

In common with all electronic components, voltage current and power levels in a BJT must not be exceeded if the device is to be operated reliably. Manufacturers' specified ratings include maximum collector and base current levels, the minimum breakdown voltages (BV_{CB} and BV_{BE}) of the collector-base and emitter-base junctions as well as a corresponding limit on collector-emitter voltage BV_{CE}.

For silicon BJTs, $|BV_{BE}|$ is usually guaranteed to be not less than $6\,V$ while $|BV_{CB}|$ can range from $10\,V$ to several hundreds of volts.

The maximum power dissipation (P_D or $P_{C(max)}$), defined as the product of collector current and collector-emitter voltage, restricts operation to an area on the output characteristic bounded by a hyperbola (Fig. 1.9). $P_{C(max)}$ is usually quoted for temperatures up to $25\,°C$. Above this temperature the allowable dissipation is reduced by a figure (in mW/°C) which depends on the thermal properties of the device package.

Integrated circuits

An integrated, or monolithic, circuit is a small chip of silicon (typically between 1 mm and 5 mm square and 0.25 mm thick) containing hundreds of thousands of components and capable of performing tasks ranging from simple combinational logic or amplification to very complex functions such as those required in microprocessors.

Morant (1990) and Chapter 5 of Sparkes (1987) are comprehensive references.

Chips are not manufactured individually but are processed, many thousands at a time, on circular silicon slices (wafers), 50 to 150 mm in diameter. When the processing and testing of wafers are complete, they are split into individual chips prior to packaging and final testing.

The planar process

The feasibility of making integrated circuits is due entirely to the planar process which, as its name suggests, involves processing only on one side of a silicon wafer. This comprises three fundamental operations – oxidation, diffusion and metallization.

Oxidation The surface of silicon is readily oxidized at elevated temperatures to form a thin insulating layer. Using photographic resist and selective etching techniques, a **window** is created in the oxide exposing the desired area of the silicon surface.

Diffusion If an impurity (usually boron for p-type, phosphorus for n-type) in gaseous form is passed over the heated silicon, it diffuses into the exposed silicon with the rest of the surface being protected by the oxide mask. Impurity diffusion occurs both vertically and laterally forming a junction under the protective oxide. Successive masking plus p-type and n-type diffusions into the silicon produce vertical diode and BJT structures.

Metallization When all diffusions are complete and the device structures have been formed, they are connected in circuit configuration by coating the surface with a thin film of aluminium and then etching away (as in printed circuit fabrication, but on a microscopic scale) all except the interconnection pattern.

Using the planar process, it is possible to fabricate many types of electronic components such as diodes, BJTs, field-effect transistors, resistors and very low-value capacitors.

Figure 1.10 illustrates the structures of a BJT, a resistor and a capacitor all isolated from each other. Isolation is essential in integrated circuits to minimize unintended interaction between components and is achieved in the following manner. The starting point is usually a p-type doped silicon wafer or substrate on which is grown an 'epitaxial' crystalline layer of n-type silicon. A deep p$^+$ diffusion through the n-type layer joins with the substrate forming n-type wells which are isolated from each other by reverse biasing their junctions with the substrate (the substrate is connected

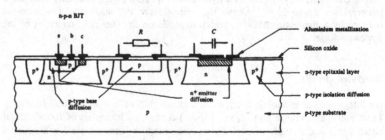

Fig. 1.10 Section (not to scale) of an isolated integrated BJT, resistor and capacitor.

to the most negative potential in the circuit). Isolated bipolar transistors, resistors and capacitors are fabricated in these wells.

Integrated BJTs

An integrated n-p-n planar BJT (see Fig. 1.10) is fabricated by performing a p-type base diffusion followed by an n^+ emitter diffusion: a combination of time, temperature and dopant concentration determines impurity profiles. The n^+ diffusion is also applied to the collector contact area since the aluminium metallization is a p-type (Group 3) impurity and, otherwise, would create an unwanted rectifying junction at the contact.

In addition to the diffusion parameters mentioned, the performance of a planar BJT is determined by its surface geometry, i.e. by the masks. If two transistors have identical geometries and are fabricated adjacent to each other (perhaps within 0.1 mm) they are subject to almost identical processing conditions and are closely matched in terms of current gain (β) and V_{BE} for a given I_C. Since the transistors are in very close thermal proximity, their parameters remain closely matched over a wide range of temperature. This feature is called (thermal) tracking. In practice, integrated BJTs have V_{BE}s matched to within 5 mV with less than 10 μV/°C drift and their βs are matched to within ± 10%.

It is rather more difficult to realize p-n-p bipolar transistors in an essentially n-p-n process which is controlled to yield adequate values of β and breakdown voltages for n-p-n devices. It is possible to use the substrate in a vertical p-n-p structure combining the substrate as collector, the n-type well as base and the p-type base diffusion as emitter. This structure has the disadvantages that the base region is rather wide, giving a low value of β, and that the substrate (the p-n-p collector) must be connected to the most negative circuit potential in order to achieve isolation for other devices.

A lateral structure for a p-n-p transistor (Fig. 1.11) can be created by diffusing both the p-type collector and emitter at the same time (the base diffusion for n-p-n devices). This lateral p-n-p transistor exhibits poor and variable performance owing to mask and processing tolerances; βs are often as little as 10. The current gain of a lateral p-n-p transistor can be enhanced by compounding it with a high-β n-p-n transistor as discussed in Chapter 4.

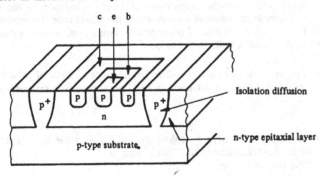

Fig. 1.11 Structure of a lateral p-n-p transistor (not to scale). The oxide layer and metallization have been omitted.

Millman and Grabel (1987), pp. 191–193 provide more information on integrated diodes.

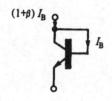

$(1+\beta) I_B$

I_B

Integrated diodes

A junction diode with a relatively high breakdown voltage (≈ 30 V) can be realized by using the collector-base junction of an integrated BJT; the emitter diffusion is unnecessary. Alternatively, the emitter-base junction ($|BV| \approx 7$ V) may be used for low-voltage or breakdown diode (Zener) applications.

Neither of these devices matches the input characteristic of a BJT which is more closely approximated by the emitter-base junction with the collector shorted to the base. This diode-connected transistor has wide application in integrated circuits both linear (see Chapter 5) and digital.

Integrated resistors

The ohmic value of an integrated resistor is realized by carefully defining the surface geometry of a base (or emitter) diffusion which has a controlled depth and resistivity. Isolation of the resistive region is provided by reverse biasing its junction with the collector well (or base region). The emitter diffusion with its low resistivity is preferred for low-value resistors ($10\,\Omega$ to $1\,k\Omega$) while the base diffusion is appropriate for higher-valued resistors (up to $50\,k\Omega$).

See Sparkes (1987), Chapter 5.

Integrated resistor values are calculated using the concept of sheet resistivity as follows. Resistive material has a bulk resistivity (ρ, in Ω-cm) which relates resistance (R) to the resistor dimensions length (ℓ), width (w) and thickness (t).

$$R = \rho \frac{\ell}{wt} \tag{1.21}$$

If a square geometry ($\ell = w$) is considered, the resistance between opposite faces of the square is

$$R = \rho \frac{\ell}{\ell t} = \frac{\rho}{t} \equiv R_S \tag{1.22}$$

where R_S is defined as the sheet resistivity (in Ω per square), independent of the size of the square.

A resistance of value ($n \times R_S$) is achieved by a surface shape which is n squares in length, an aspect ratio of n:1 (long and thin). Alternatively, a resistance of less than R_S has an aspect ratio of less than unity (short and wide). Theoretically, the actual width of the resistor is unimportant since it is only the aspect ratio that counts (for a given R_S). In practice, because of photographic limitations, resistor widths are usually made no less than 0.025 mm (1 mil).

To improve compactness, high-valued diffused resistors have a meandering or snake-like geometry. Special allowance, over and above the ohms per square, must be made for corners and end contacts.

Exercise 1.5   Determine the length of a straight, base-diffused integrated resistor of value 8 kΩ if the sheet resistivity of the base diffusion is $200\,\Omega$ per square. The resistor width is 25 μm.
[*Answer*: 1 mm]

Exercise 1.6   Calculate the aspect ratio of a base-diffused resistor of value $50\,\Omega$. The sheet resistivity of the base diffusion is $200\,\Omega$ per square.
[*Answer*: 1:4 (short and wide)]

Owing to process variations, the absolute value of an integrated resistor is subject to wide tolerance (typically up to $\pm 20\%$) but, as with integrated BJTs, resistors fabricated in close proximity to each other exhibit almost identical departures (within ± 1%) from their design values. Also, although integrated resistors vary in value with temperature (+ 0.2%/°C), physically adjacent resistors have the same temperature coefficient and are subject to the same temperature. The effect of these features is that, while the values of individual resistors are subject to temperature and tolerance variations, ratios of resistor values correspond closely to their design geometries and remain constant with temperature.

A major problem with integrated resistors is the area which they expend; a 50 kΩ resistor (base-diffused, 50 μm width) will occupy an area of 0.625 mm^2 compared with around 0.05 mm^2 for a typical low-power BJT. Imperfections in the crystal structure occur randomly over the area of a silicon wafer and, since each imperfection may result in a faulty device (and chip), the area of each chip should be minimized in order to maximize the final yield of functional chips. With chip area at a premium, total chip resistance is limited to an absolute maximum of around 500 kΩ but, more important, special circuit design techniques (see Chapter 5) are used to reduce chip resistance with scant regard for transistor count. This is a total reversal of the economics of discrete component circuit design in which the cost of a transistor is typically five to ten times that of a resistor, whatever its ohmic value.

Thin- and thick-film resistors are also used in specialized, precision circuits.

Integrated capacitors

We have seen, earlier in this chapter, that a reverse biased p–n junction exhibits a transition capacitance. This may be used in integrated circuits to realize capacitors but there are several disadvantages. The value of transition capacitance is dependent on reverse voltage and such a capacitor is polarized; the junction must be reverse biased. Also, the capacitance per unit area is very small.

An alternative approach is to create a non-polarized capacitor with electrodes formed by the low-resistivity emitter diffusion and aluminium metallization separated by a very thin (500 Å) silicon oxide dielectric (see Fig. 1.10). This structure also has a very low capacitance per unit area (in the order of 400 pF/mm^2) and, for this reason, total chip capacitance is usually limited to a maximum of approximately 100 pF. Therefore, it is impossible to integrate capacitors with nanofarad values and integrated circuit design techniques avoid, wherever possible, the use of capacitors. Essential high-valued capacitors must be discrete components external to the integrated circuit package. However, it is possible to accommodate the internal 3–10 pF compensating capacitor commonly used to stabilize operational amplifiers against oscillation.

This structure is called Metal Oxide Semiconductor (or MOS).

At this stage we are concerned primarily with bipolar techniques but it is essential to realize that MOS transistors are the key to the integration of very complex circuits. They are described in Chapter 7.

Economic forces

Despite a high utilization of computer aids, the design of an integrated circuit is a lengthy and expensive procedure. In order that the integrated circuits themselves are inexpensive, they must be mass-produced and satisfy a wide market. Although it is possible to design integrated circuits to meet almost every conceivable specification, manufactured in small numbers their cost is prohibitive, outweighing the advantages of small size, low weight, high performance and reliability. However, custom development of special integrated circuits sometimes proves cost-effective to achieve security of a product plus enhanced reliability, particularly for space and military equipment.

Commercially available integrated circuits either provide a single function to be used in large quantities, such as digital circuits (logic gates, counters, microprocessors, memory chips, etc.) and consumer circuits (such as audio amplifiers, television signal processors and electronic games circuits), or a universal function such as an operational amplifier which, with the addition of a few components external to the integrated circuit package, can be applied for many purposes in electronic circuits and equipment.

Summary

In this chapter we have attempted to cover a very large range of semiconductor device physics, a rudimentary knowledge of which is essential to the electronic circuit designer. This material has been presented at a level which, although non-rigorous, permits explanation of the operating mechanisms and terminal behaviour of semiconductor devices such as the junction diode and bipolar junction transistor (BJT) and their structures in integrated circuit form.

Using only the simple properties of n-type and p-type doped semiconductor, we have investigated the junction diode (in equilibrium, reverse bias and forward bias) and justified its I–V characteristic both in graphical form and in the diode equation representation. The exponential relationship of the diode equation readily permits calculation of voltage increments for ratio changes of current but it is rather too cumbersome a mathematical description for use in analysis and design without computer aids. In order to facilitate calculation, simple diode models have been developed (the 0.7 V model for d.c., and the dynamic slope resistance for small-signal a.c. conditions) which, although approximate, have immense value.

Diode breakdown, normally considered a detrimental feature of the device, has been applied to the design of a simple voltage stabilizer.

Operation of the BJT (as distinct from the field-effect transistor which is considered in Chapter 7) has been studied along with its terminal parameters. Graphical BJT characteristics have little more than illustrative value owing to the large spread of transistor parameters, in particular that of the common-emitter current gain (β). Regions of operation (cut-off, active, saturated and ON) have been defined; for linear amplifier applications operation should be restricted to the normal active region but, in many switching circuits (see Chapter 6), transistors are driven firmly between the extremes of ON and OFF. Ignoring leakage currents at moderate temperatures, a simple but useful d.c. model of a silicon BJT in its normal active region is a current-controlled current source ($I_C = \beta I_B$) with a base-emitter voltage (V_{BE}) of approximately 0.7 V. Small-signal a.c. models for the BJT are developed in Chapter 3.

Integrated circuits, or silicon chips, continue to make a rapidly increasing impact on electronic circuit and system design. While a brief outline of integrated circuit fabrication (the planar process) has been presented, we have concentrated more heavily on the way in which BJTs, resistors and capacitors can be realized in integrated form. Above all, we recognize the supreme matching and thermal tracking qualities of transistor βs and V_{BE}s and of resistor ratios, features which, together with economic and chip area constraints, dictate the design strategy of integrated systems.

24

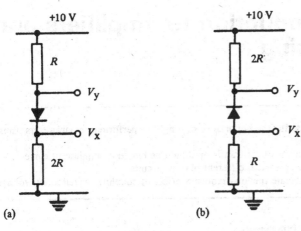

Fig. 1.12

Problems

1.1 Determine the voltages V_x and V_y in the circuits of Fig. 1.12a and b. The forward voltage of a diode is to be taken as 0.7 V.

1.2 Calculate the dynamic slope resistance at 290 K of a junction diode operating at a forward current of 2 mA.

1.3 Using the diode equation, calculate the change in voltage across a junction diode which results from a reduction in forward current to 0.25 of its initial level. Assume a temperature of 290 K.

1.4 Design a simple voltage stabilizer to provide an output voltage of 6.2 V with a load capability of 25 mA. A + 15 V supply is available and the maximum allowable dissipation of the breakdown diode is 400 mW. If the slope resistance of the breakdown diode is 7 Ω, calculate the factor by which voltage variations on the + 15 V supply will be attenuated.

1.5 The common-base current gain (α) of a BJT is measured as 0.992. Calculate the corresponding value of the common-emitter current gain (β).

1.6 The β of a BJT is measured as 300. Calculate the corresponding value of α.

1.7 Explain why the β of a BJT cannot be infinite.

1.8 Why is isolation necessary in monolithic integrated circuits?

1.9 Differentiate between a lateral p-n-p transistor and a vertical p-n-p transistor. List the advantages and disadvantages of each.

1.10 Determine the length of a straight emitter-diffused resistor of value 25 Ω if its width is 2 mil and the sheet resistivity of the emitter diffusion is 2.2 Ω per square.

1.11 Explain why a BJT should be considered as a controlled-current source rather than a controlled resistance between collector and emitter.

2 Introduction to amplifiers and biasing

Objectives

Objectives ☐ To describe the importance of amplifier performance parameters using simple models.
☐ To explain why a transistor should be biased at a quiescent current.
☐ To design several different biasing circuits.
☐ To calculate the low-frequency effect of coupling capacitors on voltage gain.

Amplifier fundamentals

An amplifier is an electronic circuit which accepts an electrical input and provides an electrical output such that there is a prescribed relationship between the input and output signals whether they be voltage or current. Normally this relationship is required to be linear so that the output is a faithful reproduction of the input. Absolute linearity is impossible to achieve even in the most carefully designed amplifiers. In practice nonlinear distortion introduced by an amplifier must be kept at an acceptable level by using techniques such as negative feedback.

Linearity is important for low distortion.

Amplifier gain

Strictly the term **gain** of an amplifier should be restricted to refer only to the ratio of output signal power to input signal power while the more specific terms **voltage gain** and **current gain** clearly relate input and output signals, both defined as voltages or as currents. The three expressions for power, voltage and current gain are dimensionless ratios since input and output signals are in the same units.

$$(\text{Power})\text{gain}, \ A = \frac{\text{Output signal power}}{\text{Input signal power}} = \frac{p_{\text{out}}}{p_{\text{in}}} \tag{2.1}$$

$$\text{Voltage gain}, \ A_{\text{v}} = \frac{\text{Output signal voltage}}{\text{Input signal voltage}} = \frac{v_{\text{out}}}{v_{\text{in}}} \tag{2.2}$$

$$\text{Current gain}, \ A_{\text{I}} = \frac{\text{Output signal current}}{\text{Input signal current}} = \frac{i_{\text{out}}}{i_{\text{in}}} \tag{2.3}$$

There are cases where the input and output signals of amplifiers are not in the same units; for example, the input may be in the form of a current (say, from a photodiode) and the amplifier generates a proportional output voltage. In this case the gain of the amplifier is the ratio of output voltage to input current, an expression which has dimensions of resistance and is defined as the **transfer resistance** (or transresistance), symbol r_{T}. Conversely the gain of an amplifier which converts voltage to current would be specified in terms of the ratio of output current to input voltage – a transfer conductance (or transconductance) g_{T}.

26

$$\text{Transfer resistance, } r_T = \frac{\text{Output signal voltage}}{\text{Input signal current}} = \frac{v_{\text{out}}}{i_{\text{in}}} \; \Omega \tag{2.4}$$

$$\text{Transfer conductance, } g_T = \frac{\text{Output signal current}}{\text{Input signal voltage}} = \frac{i_{\text{out}}}{v_{\text{in}}} \; \text{S} \tag{2.5}$$

(At low frequencies the series inductance and shunt capacitance of all circuit components may generally be neglected; series impedances are dominated by their resistances, shunt admittances by their conductances.)

Input and output resistances

The input pair of terminals (or port) of an amplifier may be represented by an input resistance (r_{in}) which, unfortunately, is usually dependent on highly variable transistor parameters such as β. A brief analysis shows, for practical input voltage and current sources (respectively represented by their Thévenin and Norton equivalent circuits incorporating a fixed source resistance R_S), that opposite constraints exist on r_{in} for the well defined transfer of voltage or current to the amplifier input.

Simple models are used to represent amplifier performance parameters.

Horrocks (1990), pp. 21–29 covers amplifier modelling in more detail.

For the voltage source of Fig. 2.1a

$$v_{\text{in}} = v_s \frac{r_{\text{in}}}{R_S + r_{\text{in}}} \tag{2.6}$$

Hence, to achieve a well defined (and maximum) voltage transfer, r_{in} must be very much greater than R_S; ideally, r_{in} should be infinite. Conversely, in the case of current drive (Fig. 2.1b)

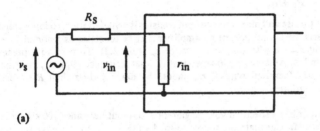

(a)

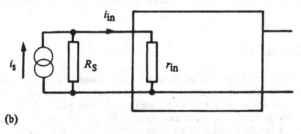

(b)

Fig. 2.1 Amplifier with (a) input voltage source and (b) input current source.

27

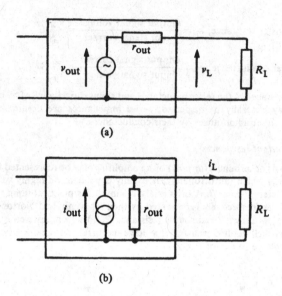

(a)

(b)

Fig. 2.2 Amplifier with (a) output voltage source and (b) output current source.

$$i_{in} = i_s \frac{R_S}{R_S + r_{in}} \tag{2.7}$$

and both precise and maximum current transfer is achieved for zero input resistance.

Similarly the output port of an amplifier may be modelled as a non-ideal voltage or current source with output resistance r_{out} (Fig. 2.2). To achieve a precise and maximum load voltage (v_L) independent of load resistance (R_L), r_{out} must be zero. In the case of current output, r_{out} should be much greater than R_L; ideally r_{out} should be infinite.

Exercise 2.1 An amplifier has a specified voltage gain of 100, input resistance 100 kΩ and output resistance 10 Ω. Calculate the source-to-load voltage gain for a source resistance of 10 kΩ and load resistance 100 Ω. [*Answer*: $A_v = 82.6$]

Compare your result with the ideal case of zero source resistance and infinite load resistance.

BJT configurations

Bipolar junction transistors (BJTs) are widely used as amplifying devices and, since they are three-terminal devices, there are several ways of determining which pair of terminals is the input port and which pair the output port. The three useful configurations (in which the input can control the output) shown in Fig. 2.3 are named by the terminal which is common to both input and output ports: (a) common-emitter, (b) common-base and (c) common-collector (or emitter follower).

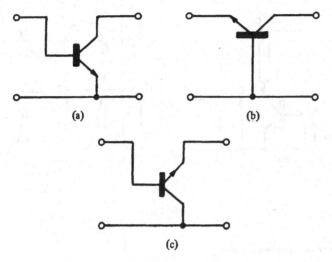

(a)
(b)
(c)

Fig. 2.3 BJT configurations: (a) common-emitter, (b) common-base, (c) common-collector.

Later they are considered in detail with respect to their terminal properties, i.e. transfer parameter, input resistance and output resistance.

BJT biasing in common-emitter

The need for biasing

Consider the common-emitter amplifier of Fig. 2.4 which includes a collector resistor (R_C) to provide a path for the collector current and to convert the collector current changes to an output voltage signal.

If a sinusoidal input voltage (V_{in}), alternately positive and negative with respect to earth, is applied directly to the n–p–n transistor base, the transistor conducts on positive half-cycles and cuts off on the negative excursions. In order to avoid this distortion of the output, it is essential to bias the transistor to a finite quiescent collector current (and corresponding output voltage) and to superimpose the signal variations on the bias level.

The transistor must be biased to allow bidirectional swings.

To obtain maximum output voltage swing, the collector is biased half way between its extreme allowable excursions, i.e. cut-off, when no collector current flows, and saturation, when the collector voltage falls to meet the base voltage (the limit of collector-base reverse bias).

Adding extra components to the basic circuit allows us to establish a nominal value of quiescent collector current. With ideal devices showing no variation in β, V_{BE} or leakage currents, bias circuits would be accurate and stable over wide ranges of temperature and independent of transistor substitution. Detailed analysis of practical devices is tedious and of value only when precise measures of quiescent collector current and its variability are required. The design approach for several bias circuits which are in common usage is considered in the following sections.

Millman and Grabel (1987), Chapter 11 treat the stability of bias circuits in some detail.

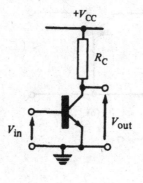

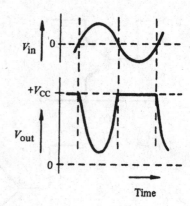

Fig. 2.4

Constant base current biasing

The simplest bias circuit is shown in Fig. 2.5. Resistor R_B provides the appropriate base current for the transistor.

This allows approximately equal positive- and negative-going collector voltage excursions about $0.5 V_{CC}$.

Assuming that a quiescent collector voltage of $0.5 V_{CC}$ is specified,

$$\text{Quiescent } I_c = \frac{V_{CC}}{2R_C} \tag{2.8}$$

and

$$\text{Quiescent } I_B = \frac{1}{\beta} \frac{V_{CC}}{2R_C} \tag{2.9}$$

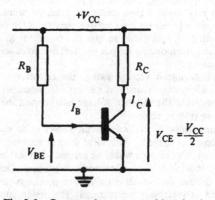

Fig. 2.5   Constant base current bias circuit.

R_B can now be calculated by

$$R_B = \frac{(V_{CC} - V_{BE})}{I_B} \tag{2.10}$$

$$R_B = (V_{CC} - V_{BE})\frac{2\beta R_C}{V_{CC}} \tag{2.11}$$

The base current is well defined since V_{BE} stays substantially constant at approximately 0.7 V, but note that R_B must be selected on test since its design dependence on β gives it a range of typically 10:1 in value!

Not a very useful circuit in practice as proved by the design example.

Design Example 2.1

Given a +12 V power supply, bias an n-p-n transistor using constant base current bias to quiescent conditions: $V_{CE} = +6$ V, $I_C = 4$ mA. The transistor type has minimum, typical and maximum β values of 50, 150 and 300 respectively and a V_{BE} of 0.7 V may be assumed. Determine the suitability of your design for devices at the limits of the β spread.

Solution. Referring to Fig. 2.5,

$$R_C = \frac{V_{CC} - V_{CE}}{I_C}$$

$$= \frac{12 - 6}{0.004}$$

$$= 1.5 \text{ k}\Omega \text{ (preferred value)}$$

For a program which performs constant base current bias calculations, see Attikiouzel and Jones (1991), pp. 50–54.

$$R_B = (V_{CC} - V_{BE})\frac{\beta}{I_C}$$

$$= \frac{11.3 \times 150}{0.004} = 423.8 \text{ k}\Omega$$

for the typical β value of 150.

Choose R_B as 390 kΩ, which is the nearest preferred value. I_C should be calculated for this R_B

Nearest preferred values of components must be used in practical circuit design. Here we use the E12 series of preferred values (see Appendix A).

$$I_C = \beta\frac{V_{CC} - V_{BE}}{R_B} = 4.35 \text{ mA}$$

and

$$V_{CE} = V_{CC} - I_C R_C$$

$$= 5.48 \text{ V}$$

Both I_C and V_{CE} are reasonable approximations to the desired bias levels considering that we have taken a nearest preferred value for R_B. Now what happens if we use transistors of minimum and maximum β? With the minimum β of 50,

$$I_C = 1.45 \text{ mA}$$

and

$$V_{CE} = 9.83 \text{ V}$$

differing substantially from our design target and restricting the bidirectional output voltage swing to some 2 V peak, and for the maximum β (300),

$$I_C = 8.69 \, \text{mA}$$

and

$$I_C R_C = 13.04 \, \text{V}$$

which is greater than V_{CC}! This calculation tells us that, for a β of 300, such a transistor would be saturated and no negative-going output signal swings could be handled.

Shunt feedback biasing

Leakage currents usually can be ignored for laboratory circuits but should be taken into consideration for circuits designed to operate at elevated temperatures.

Much more satisfactory biasing can be achieved by providing as feedback a measure of the collector current (and collector voltage), thus reducing the effect of β and V_{BE} spreads and temperature effects on V_{BE} and leakage currents. Such feedback can be taken either from the collector or from a resistor introduced in the emitter lead of the transistor.

Referring to Fig. 2.6 in which R_B is now taken to the collector rather than to the supply,

$$R_B = (0.5V_{CC} - V_{BE})\frac{2\beta R_C}{V_{CC}} \tag{2.12}$$

for $V_{CE} = 0.5V_{CC}$. If $0.5V_{CC} \gg V_{BE}$, $\tag{2.13}$

$$R_B = \beta R_C$$

Horrocks (1990), Chapter 4 presents a formal treatment of feedback amplifier configurations. This circuit is more correctly termed 'shunt *voltage* feedback'.

Although this does not exhibit an obvious decrease in β dependence, the feedback does tend to stabilize the operating point firmly within the limits of saturation and cut-off. For example, if the transistor has a much higher β than the nominal value used in calculating R_B, the collector current is greater than (and the collector voltage less than) the design value. With the lower collector voltage, however, there

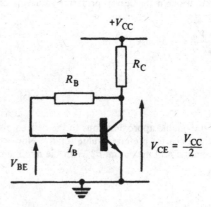

Fig. 2.6 Shunt feedback bias circuit.

is less voltage across R_B and therefore less than the nominal base current, thus compensating (partially) for the high β. Conversely, if β is lower than nominal, the collector current is less than the design value giving a greater voltage across R_B and more base current, again partially compensating for the low β.

With R_B connected in this feedback position, both d.c. (bias) and a.c. (signal) feedback occur. The former is desirable to stabilize the d.c. operating point, whereas signal feedback may not be desirable. If this is the case, the a.c. feedback may be removed, while still retaining the d.c. feedback, by splitting R_B into two series resistors and decoupling their midpoint to earth with a capacitor.

Note the stabilizing action of negative feedback.

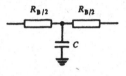

Use shunt biasing to realize the specification of Design Example 2.1.

Design Example 2.2

Solution. Referring to Fig. 2.6,

$$R_C \approx \frac{V_{CC} - V_{CE}}{I_C}$$

$$= \frac{12 - 6}{0.004}$$

$$= 1.5\,\text{k}\Omega \ (\text{preferred value})$$

and

$$R_B = (V_{CE} - V_{BE})\frac{\beta}{I_C}$$

$$= \frac{5.3 \times 150}{0.004}$$

$$= 198.8\,\text{k}\Omega$$

Use the nearest preferred value of 180 kΩ for R_B. It is necessary to analyse the circuit to obtain expressions relating the bias V_{CE} and I_C to the known circuit parameters. The results are quoted but the reader should justify them.

$$V_{CE} = \frac{V_{CC}R_B + (1 + \beta)V_{BE}R_C}{R_B + (1 + \beta)R_C}$$

and

$$I_C = \beta\frac{V_{CE} - V_{BE}}{R_B}$$

Substituting for the typical β device gives

$$V_{CE} = 5.7\,\text{V}$$

$$I_C = 4.17\,\text{mA}$$

For the minimum β of 50,

$$V_{CE} = 8.63\,\text{V}$$

$$I_C = 2.2\,\text{mA}$$

and for a β of 300,

$$V_{CE} = 3.92\,V$$

$$I_C = 5.37\,mA$$

Tabulate the above results against those obtained in Design Example 2.1. Comparison clearly shows the advantage of d.c. negative feedback.

Potentiometer biasing

A third method of biasing is shown in the circuit of Fig. 2.7 where a resistor (R_E) is introduced into the emitter circuit.

A useful and generally accurate approximation.

$$V_E \approx I_C R_E \text{ (since } I_C \approx I_E) \tag{2.14}$$

and

$$V_B = V_E + V_{BE} \tag{2.15}$$

$$\approx I_C R_E + V_{BE} \tag{2.16}$$

V_{BE} spread and temperature variations affect the bias.

Hence if V_B were fixed (a d.c. voltage source), it and R_E would determine I_E and I_C independent of β and the base current. However, in practice, there is a spread of approximately $\pm 50\,mV$ on V_{BE} for a defined operating current and a temperature coefficient of approximately $-2\,mV/°C$; thus I_E is not absolutely fixed. The variability of I_E is greatly reduced as V_B is increased; the variation in V_E (which is proportional to I_E) can thus be small. It is usual to allow V_E to be in the range 1–3 V, i.e. much larger than the possible V_{BE} variation and the larger the better.

Note, again, the stabilizing action of negative feedback which is series *current* feedback (Horrocks, 1990, p. 61).

The action of the feedback introduced by R_E is described as follows. Assuming that the collector current tries to increase (owing to a rise in temperature, say), V_E increases and, with V_B fixed, V_{BE} is forced to fall. This reduction in V_{BE} causes the collector current to reduce again, close to its nominal value, thus partially compensating for the initial increase.

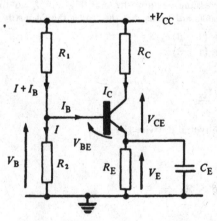

Fig. 2.7   Potentiometer bias circuit.

V_B cannot be realized as a perfect voltage source since its a.c. short-circuit property would divert the input signal current away from the transistor base and no output signal would be present. It is customary to derive V_B using resistors R_1 and R_2 as a potential divider (or potentiometer) with an effective resistance $R_B = R_1 \| R_2$. The lower the value of R_B, the less the resulting variation in I_C owing to β spread. R_1 and R_2 are usually chosen on the basis of allowing a current of ten times maximum base current (corresponding to $I_C / \beta_{\min}$) to flow through R_1 and R_2.

To preserve a common-emitter configuration without a.c. feedback, the emitter voltage must not be allowed to change in sympathy with applied a.c. signals. The emitter is therefore decoupled with a high-valued capacitor (C_E), usually electrolytic for audio frequencies.

This biasing technique is also known as **self-bias** or **emitter bias**.

Design Example 2.3

Using potentiometer biasing, establish quiescent I_C and V_{CE} of 4 mA and 6 V respectively in a common-emitter amplifier using the n-p-n transistor specified in Design Example 1.

Solution. Let $V_E \approx 3$ V. Then

$$R_E = \frac{V_E}{I_C} = \frac{3}{0.004} = 750\,\Omega$$

Use $R_E = 680\,\Omega$ (nearest preferred value) and decouple with a high-valued capacitor for common-emitter operation. Calculating V_E for this value of R_E and required I_C gives

$$V_E = I_C R_E = 0.004 \times 680 = 2.72\,\text{V}$$

and

$$V_B = V_E + V_{BE} \approx 2.72 + 0.7 = 3.42\,\text{V}$$

Now for maximum bidirectional signal swing,

$$I_C R_C = V_{CE} = 6\,\text{V}$$

Therefore

$$
\begin{aligned}
V_{CC} &= V_E + V_{CE} + I_C R_C \\
&= 2.72 + 6 + 6 \\
&= 14.72\,\text{V}
\end{aligned}
$$

A collector supply voltage (V_{CC}) of +15 V would be used in practice.

$$R_C = \frac{6}{0.004} = 1.5\,\text{k}\Omega \text{ (a preferred value)}$$

The quiescent V_{CE} may now be checked by

$$
\begin{aligned}
V_{CE} &= V_{CC} - I_C R_C - V_E \\
&= 15 - (0.004 \times 1500) - 2.72 \\
&= 6.28\,\text{V}
\end{aligned}
$$

Since the transistor has a minimum β of 50,

$$I_{B(max)} = \frac{4}{50} \text{ mA}$$

$$= 80 \, \mu A$$

Referring to Fig. 2.7, let $I \approx 1 \text{ mA}$ (approx 10 times $I_{B(max)}$).

$$R_2 = \frac{V_B}{I} = \frac{3.42}{0.001} = 3.42 \text{ k}\Omega$$

Use $R_2 = 3.3 \text{ k}\Omega$ (nearest preferred value) and recalculate I,

$$I = \frac{V_B}{R_2} = \frac{3.42}{3300} = 1.036 \text{ mA}$$

Assuming a typical β of 150,

$$I + I_{B(typ)} = 1.036 + 0.027 \text{ mA}$$

$$= 1.063 \text{ mA}$$

Therefore

$$R_1 = \frac{15 - 3.42}{1.063} \text{ k}\Omega$$

$$= 10.89 \text{ k}\Omega$$

Note the methodical step-by-step approach to the design procedure and iteration when nearest preferred values are used.

The preferred value of $10 \text{ k}\Omega$ is selected for R_1.
This completes the design, the results of which may be summarized:

$V_{CC} = +15 \text{ V}$	$R_1 = 10 \text{ k}\Omega$
$R_C = 1.5 \text{ k}\Omega$	$R_2 = 3.3 \text{ k}\Omega$
$R_E = 680 \, \Omega$	C_E is large

Biasing other configurations

Although the potentiometer bias circuit is emitter-decoupled to yield a common-emitter amplifier, the basic circuit is readily adaptable to both common-base and common-collector configurations (Fig. 2.8).

To create the common-base configuration (i.e. earthed base for a.c. signals), connecting a capacitor between base and earth is all that is necessary. In common-collector operation, the collector supply itself is a short-circuit to a.c. signals and simply omitting the collector load resistor yields the desired circuit.

Coupling capacitors

Up to this stage we have considered in isolation the problem of designing a bias circuit. We have ignored the quiescent or reference voltage of an input signal source and any external load that may be connected to the output of the circuit. Indeed any d.c. path through source or load may severely complicate the design. A solution to this problem is to isolate both source and load from the amplifier by d.c. blocking

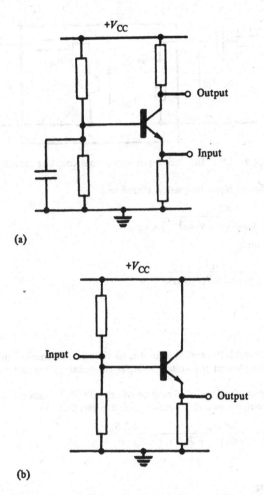

(a)

(b)

Fig. 2.8 Potentiometer bias applied to (a) common-base and (b) common-collector configurations.

(or coupling) capacitors which, provided they are of sufficiently high value, permit the unattenuated transfer of a.c. signals but avoid upsetting the d.c. conditions.

These capacitors influence the transfer of low-frequency a.c. signals as is demonstrated by analysis of the amplifier model shown in Fig. 2.9.

At the input

$$v_{in} = v_s \frac{r_{in}}{R_S + r_{in} + \dfrac{1}{j\omega C_S}} \qquad (2.17)$$

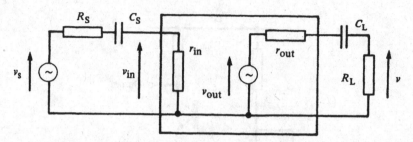

Fig. 2.9 Amplifier with input and output coupling capacitors.

where ω = angular input frequency, therefore

$$v_{in} = v_s \frac{j\omega C_S r_{in}}{1 + j\omega C_S (r_{in} + R_S)} \tag{2.18}$$

and at the output

$$v_L = v_{out} \frac{j\omega C_L R_L}{1 + j\omega C_L (R_L + r_{out})} \tag{2.19}$$

also

$$A_v = \frac{v_{out}}{v_{in}} \tag{2.20}$$

Equations 2.18 and 2.19 show that for d.c. ($\omega = 0$) signal transmission is zero while at very high frequencies the voltage transfer relationships are identical to that of Equation 2.6.

Also, Equations 2.18 and 2.19 may be combined with Equation 2.20 to yield the rather more complex overall source-to-load transfer function

$$\frac{v_L}{v_s} = A_v \frac{j\omega C_S r_{in}}{1 + j\omega C_S (r_{in} + R_S)} \frac{j\omega C_L R_L}{1 + j\omega C_L (R_L + r_{out})} \tag{2.21}$$

Horrocks (1990), pp. 29–39 covers frequency response of amplifiers in more detail.

Direct coupling

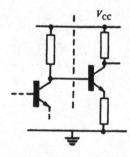

It is impossible to incorporate high-valued capacitors in integrated circuits intended for use at low frequencies. If essential they are included as components external to the integrated circuit package but by a useful design technique may be dispensed with altogether. The technique is known as direct coupling.

By suitable choice of quiescent voltage levels in the circuit, a transistor amplifier input may be connected directly to the output of the preceding stage. A d.c. path is thus provided for both quiescent base current and input signal with none of the low-frequency response degradation associated with a coupling capacitor.

It should be noted that many electronic systems require multiple power supplies, e.g. +15 V and −15 V, when operational amplifiers are used. It is possible, and indeed desirable, to make use of a negative supply when biasing a transistor. Consider the case where the n–p–n transistor emitter is connected to −15 V via emitter resistor R_E. Owing to the large voltage drop across R_E (≈ 14.3 V if $V_{in} = 0$ V), stability of the quiescent collector current is greatly improved compared with a circuit in which only a 2 V drop has been allocated. An emitter decoupling capacitor (C_E) must still be retained, however, to preserve common-emitter action.

Amplifier designs which dispense with capacitors completely rely on rather more complex multiple-transistor circuits such as the differential amplifier, current mirror and so on, which are introduced in Chapter 5.

It is important to note that, while we have considered n–p–n transistors in all the examples, the techniques presented are identical for p–n–p transistors, only all voltage polarities and current directions should be reversed.

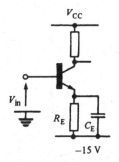

What about p–n–p transistors?

Summary

An amplifier stage is usually only a small part of a complex electronic system and, since it may be a complicated circuit in its own right, the ability to recognize its important system parameters is essential. Use of the simple model incorporating an input-controlled output source plus input and output resistances allows us to determine its performance within the system.

Before considering the circuit behaviour of the three viable BJT configurations, it is recognized that a transistor must be biased to a quiescent conduction level to allow processing of bidirectional signals. Several widely used bias circuits have been treated from a design viewpoint but the underlying value of negative feedback techniques should be appreciated.

Particular attention should be paid to the step-by-step nature of design procedures, each part aiming at yielding one component value. It is seldom that procedures involving the analysis of complete circuits and solving three or more simultaneous equations are of value. Also, note the assignment of nearest preferred values as the design proceeds rather than completing a design and then rounding the component values.

The single-transistor amplifier stage must not be considered in isolation; input and output coupling techniques can affect its performance, particularly at low frequencies where the reactance of coupling capacitors cannot be ignored.

Problems

2.1 Determine the minimum input resistance required for an amplifier driven from a source resistance of 1 kΩ such that the input voltage attenuation is less than 2%.

2.2 Determine the maximum output resistance required for an amplifier driving an external load of 10 kΩ such that the output voltage attenuation is less than 1%.

2.3 Calculate the quiescent I_C and V_{CE} of the transistor in the constant base current bias circuit of Fig. 2.5 given that $V_{CC} = +9$ V, $R_B = 270$ kΩ and $R_C = 1$ kΩ. The β of the transistor is 100 and its V_{BE} may be assumed to be 0.7 V.

2.4 If the base resistor (R_B) in Problem 2.3 is now connected between collector and base of the transistor (shunt feedback bias), calculate the new d.c. values of I_C and V_{CE}.

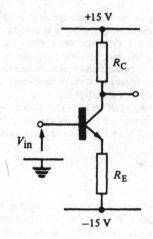

+15 V

R_C

V_{in}

R_E

−15 V

Fig. 2.10

2.5 With reference to Fig. 2.10, if $R_E = 3.3\,k\Omega$ and $R_C = 1.5\,k\Omega$, calculate the quiescent I_C and V_{CE} for $V_{in} = 0\,V$.

2.6 Repeat Problem 2.5 for $V_{in} = +2\,V$.

2.7 Using a shunt feedback circuit, bias an n-p-n transistor to quiescent conditions of $I_C = 1\,mA$ and $V_{CE} \approx +6\,V$. A $+15\,V$ supply is available and the transistor may be assumed to have a β of between 80 and 320.

2.8 Figure 2.11 shows a common-emitter amplifier. Calculate
(a) the voltages at the collector, base and emitter under no-signal conditions, and

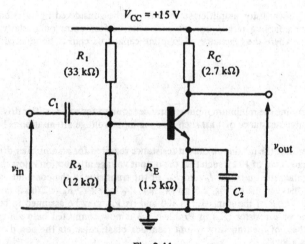

$V_{CC} = +15\,V$

R_1
(33 kΩ)

R_C
(2.7 kΩ)

C_1

v_{out}

v_{in}

R_2
(12 kΩ)

R_E
(1.5 kΩ)

C_2

Fig. 2.11

(b) the range of values over which the instantaneous collector and emitter voltages can be varied when an a.c. input signal (v_{in}) is applied.

2.9 List and verify the design procedures used to select R_1, R_2 and R_E in the circuit of Fig. 2.11.

2.10 Repeat Problem 2.7 using a potentiometer bias circuit.

2.11 Repeat Problem 2.8a with $R_1 = 14.6\,\text{k}\Omega$, $R_2 = 5.4\,\text{k}\Omega$, $R_C = 1\,\text{k}\Omega$, $R_E = 500\,\Omega$ and $V_{CC} = +10\,\text{V}$.

2.12 The amplifier shown in Fig. 2.12 has been constructed. On testing, it is found that the d.c. collector voltage is $+8.6\,\text{V}$ instead of the specified value of $10.6\,\text{V}$. This fault is due to one resistor being short-circuit: which resistor is it?

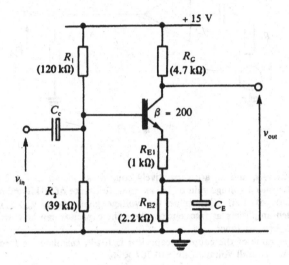

Fig. 2.12

2.13 The amplifier of Fig. 2.13 has the following parameter values: input resistance, $r_{in} = 10\,k\Omega$; output resistance, $r_0 = 20\,\Omega$; transconductance, $g_T = 50\,S$. The source and load resistances are $R_s = 1\,k\Omega$ and $R_L = 100\,\Omega$ respectively. Calculate the voltage gain, A_V ($= v_0/v_s$), at frequencies where the capacitor can be considered as having zero reactance.

If the value of the coupling capacitor is $4.7\,\mu F$, calculate the frequency at which the voltage gain is $0.707 \times A_V$.

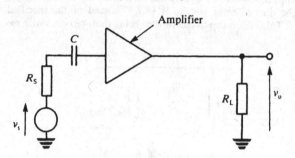

Fig. 2.13

2.14 Amplifiers A_1 and A_2 are capacitively coupled as shown in Fig. 2.14. Each amplifier has a voltage gain of 10, an input resistance of $10\,k\Omega$ and an output resistance of $1\,k\Omega$. Determine the voltage gain, A_V ($=v_{out}/v_{in}$), of the cascaded amplifiers at frequencies where the capacitor can be considered as having zero reactance.

If the value of the coupling capacitor is $10\,\mu F$, calculate the frequency at which the overall voltage gain is $0.707 \times A_V$.

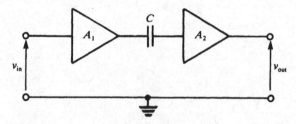

Fig. 2.14

Small-signal BJT models and their application 3

Objectives

- ☐ To develop a simple BJT model valid for small signals.
- ☐ To explain the construction of an a.c. equivalent circuit of an amplifier from its full circuit diagram.
- ☐ To analyse the equivalent circuit (incorporating a simple model for the BJT) and calculate gain, input resistance and output resistance.
- ☐ To recognize (but not use) the full hybrid-π model.

Why model?

In Chapter 2 we discussed how to bias a BJT in the three circuit configurations: common-emitter, common-base and common-collector. This was essentially a d.c. consideration and provided no information as to how the biased transistor behaves as an amplifier. It is important to be able to quantify the a.c. signal behaviour of the amplifier in terms of gain, input resistance and output resistance. Results of a.c. analysis are used to design a circuit; the initial choice of circuit configuration to meet a specification is based on experience of the performance of different circuits.

It is possible to use the graphical characteristics of a transistor (introduced in Chapter 1) to calculate the behaviour of the device in circuit; for example, the input characteristic (I_B versus V_{BE}) gives the $\triangle I_B$ for a given $\triangle V_{BE}$ (input voltage change in common-emitter). The transfer characteristic (I_C versus I_B) can then be used to calculate the corresponding $\triangle I_C$ and $\triangle V_{out} (= \triangle I_C \times R_C)$. Figures for the voltage gain ($\triangle V_{out} / \triangle V_{BE}$) and input resistance ($\triangle V_{BE} / \triangle I_B$) then follow. Using this method, however, there are three major problems:

1. The slope (β) of the transfer characteristic can vary typically over a 10:1 range. The question arises of whether the β of the device should be the typical, minimum or maximum value when performing the analysis.
2. The input characteristic is exponential in shape and actual increments of V_{BE} and I_B must be taken when considering large signals. The tangential slope of the characteristic at the quiescent bias point is valid only for small-signal calculations.
3. Manufacturers do not always provide graphical characteristics in their data for BJTs. The transfer and output characteristics are so β-dependent as to make them almost worthless for analysis.

The alternative approach is to use a model (equations or equivalent circuit) which describes the behaviour of the device. Considering large signals, where nonlinearity of the transistor is significant, usual representation is by the Ebers-Moll equations (beyond the scope of this text) but for small signals a BJT can be represented by

Graphical methods can be very useful if the characteristics relate to your particular device; otherwise they serve only to illustrate principles.

See Sparkes (1987), Chapter 4.

The Ebers-Moll equations and their extension into the Gummel-Poon model are widely used in computer simulation of large-signal transistor circuits by, for example, the SPICE package. See Sparkes (1987), Chapters 2 and 4, and Morant (1990), Chapter 9.

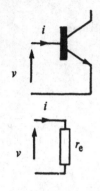

circuit components – slope resistances, capacitors and a current source which reflect the terminal behaviour of the device taking into account β spread and bias dependence.

A basic BJT model

A small-signal model for a BJT can be developed from the a.c. diode model presented in Chapter 1. Consider a BJT with its collector open-circuit. The small-signal behaviour of the forward biased base-emitter diode can be represented by its slope resistance

$$r_d = \frac{v}{i} = \frac{kT}{qI_{(d.c.)}} \qquad (3.1)$$

Since this is now associated with the emitter of the BJT, it is changed to

$$r_e = \frac{kT}{qI_E} \approx \frac{25}{I_{E(mA)}} \quad \text{at } 290\,\text{K} \qquad (3.2)$$

in order to distinguish it from a simple two-terminal diode.

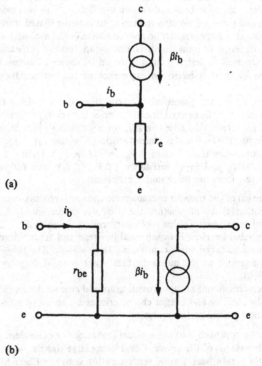

(a)

(b)

Fig. 3.1 Simple common-emitter BJT models.

From the output and transfer characteristics we recognize that the collector of a BJT is a current source; $I_C (= \beta I_B)$ is ideally independent of V_{CE} and, for a.c. signals, $i_c = \beta i_b$. Hence, when the collector is brought into circuit the model becomes that of Fig. 3.1a. The collector current source is controlled by the input small-signal base current and flows, with base current, into the emitter.

This model can be used without modification but is more amenable to analysis if its input and output sections are separated, as follows.

Considering the circuit of Fig. 3.1a, the input resistance is given by

$$r_{in} = \frac{v_{in}}{i_{in}} = \frac{(i_b + \beta i_b) r_e}{i_b} = (1 + \beta) r_e \tag{3.3}$$

Figure 3.1b shows the resistance r_{be} connected between the base and emitter terminals. Here, the input resistance is simply r_{be}. The collector current (βi_b) is the same in each case but the two models are equivalent only if

$$r_{be} = (1 + \beta) r_e \tag{3.4}$$

This second representation (Fig. 3.1b) is the one which is favoured for common-emitter modelling.

Use of the model

Consider the common-emitter amplifier of Fig. 3.2a. In this circuit the transistor has been biased to a quiescent I_C and V_{CE} and an input signal v_{in}, a.c. coupled to the base, drives the transistor around its quiescent d.c. operating point. For a.c. analysis of the circuit some components are shown to be redundant since they are incorporated only for biasing reasons.

At frequencies of interest the coupling and decoupling capacitors (C_C and C_E) are assumed sufficiently high-valued to have negligible reactance and can be replaced, for a.c. signals, by short-circuits (Fig. 3.2b).

The collector supply (V_{CC}) is a constant d.c. voltage source with zero resistance and therefore can be considered, from an a.c. viewpoint, to be connected directly to the earth reference as shown by the dotted line in Fig. 3.2b. (Alternatively, total circuit behaviour is the superposition of d.c. and a.c. conditions; for the a.c. circuit all d.c. voltage supplies are set to zero (replaced with short-circuits) and d.c. current sources removed (open circuits). Hence, for a.c. analysis, the V_{CC} supply is equivalent to earth. Similarly, the d.c. bias can be considered in isolation by setting to zero all a.c. voltage sources and removing all a.c. current sources.)

Finally, in constructing the a.c. equivalent circuit, it remains to replace the transistor symbol with its small-signal model as shown in Fig. 3.2c. It is important that voltages and currents in the a.c. equivalent circuit are labelled with care, preserving their defined directions.

Now the circuit can be analysed, as follows

$$v_{out} = - \beta i_b R_C \tag{3.5}$$

and

$$i_b = \frac{v_{in}}{r_{be}} \tag{3.6}$$

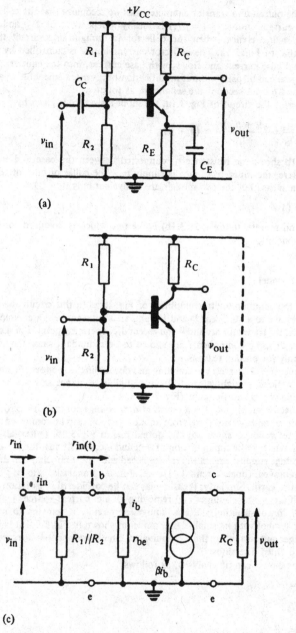

(a)

(b)

(c)

Fig. 3.2 Common-emitter amplifier and development of its a.c. equivalent circuit.

Substituting for i_b in Equation 3.5 to give v_{out} in terms of v_{in},

$$v_{out} = -\frac{\beta R_C v_{in}}{r_{be}}$$

Therefore voltage gain,

$$A_V = \frac{v_{out}}{v_{in}}$$

$$= -\frac{\beta R_C}{r_{be}}$$

$$= -\frac{\beta R_C}{(1+\beta)r_e} \tag{3.7}$$

$$= -\frac{R_C}{r_e} \quad \text{if } \beta \gg 1 \tag{3.8}$$

(The minus sign in the expression for voltage gain indicates inversion, i.e. if the input voltage rises, the output voltage falls and vice versa.) Also,

$$r_{in} = \frac{v_{in}}{i_{in}} = R_1 \| R_2 \| r_{be} \tag{3.9}$$

Sometimes erroneously referred to as 180° phase shift. This is correct only if the waveforms are sinusoidal, squarewaves or of similar symmetrical shape.

Two components of r_{in} may be identified as the equivalent Thévenin resistance of the biasing resistors, in parallel with the input resistance ($r_{in(t)}$) looking into the transistor base, where

$$r_{in(t)} = \frac{v_{in}}{i_b} = r_{be} = (1+\beta)r_e \tag{3.10}$$

Note that both A_V and r_{in} depend on r_e and consequently on the d.c. bias current ($I_E \approx I_C$) as shown by Equation 3.2; increasing I_C, A_V rises while r_{in} falls. The dependence of r_{in} on β is implicit.

The procedure for determining the output resistance (r_{out}) of a circuit is to short-circuit the input voltage source (or open-circuit an input current source) and to apply externally a voltage (v_o) across the output terminals. If the resultant current (i_o) flowing into the output is calculated, the output resistance is given by the ratio

$$r_{out} = \frac{v_o}{i_o} \tag{3.11}$$

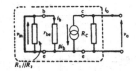

For this circuit with $v_{in} = 0$, $i_b = 0$ and the current generator βi_b is also zero, therefore

$$v_o = i_o R_C$$

and

$$r_{out} = R_C \tag{3.12}$$

An alternative method for determining output resistance is given in Fidler and Ibbotson (1989), pp. 154–158.

r_{out} is the output resistance of the circuit, i.e. R_C in parallel with the ideally infinite resistance of the collector current generator.

This application of the simple model shows that the small-signal performance of the common-emitter amplifier, in terms of A_V, r_{in} and r_{out}, can be calculated from a knowledge of β, the d.c. bias current and circuit components R_1, R_2 and R_C. In Chapters 4 and 5, the simple model is used to analyse other circuits.

Exercise 3.1 In a common-emitter amplifier circuit, the transistor has a collector load of 4.7 kΩ and is biased at a quiescent collector current of 1 mA. If the β of the transistor is 50 calculate the voltage gain and input resistance.
[*Answer*: $A_V = -184$, $r_{in} = 1.275$ kΩ]

Exercise 3.2 Repeat Exercise 3.1 for a transistor β of 100. What do you infer?
[*Answer*: $A_V = -186$, $r_{in} = 2.55$ kΩ]

Exercise 3.3 Calculate the voltage gain (v_{out}/v_s) for the circuit of Exercise 3.1 modified by the inclusion of an input source resistance of 600 Ω. Consider the cases when $\beta = 50$ and $\beta = 100$.
[*Answer*: -125, -150]

Full hybrid-π model

Sparkes, Chapters 1 and 4.

The basic BJT model incorporating r_{be} and βi_b is adequate for most low-frequency amplifier circuits; its very simplicity permits rapid analysis and yields meaningful, if only approximate, results. However, several important physical features of BJT behaviour have been omitted and need to be considered for more accurate representation. The resulting model is called the hybrid-π equivalent circuit.

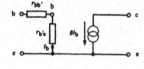

In a BJT structure the active base-emitter junction area is separated from the actual base terminal by (resistive) bulk semiconductor material. This extrinsic base resistance can be incorporated in the model as a resistance $r_{bb'}$ between the external base terminal b and the active base b'. Note that in this simplified model $r_{bb'}$ is ideally a fixed resistance independent of device current and not a slope resistance. The base-emitter slope resistance r_{be} now becomes $r_{b'e}(= (1 + \beta)r_e)$ and forms an input voltage attenuator with $r_{bb'}$, which typically has a value of between 50 and 300 Ω.

The simple model does not take into account the slope of the output characteristic which occurs in practice (called the Early effect and is due to base-width modulation). Two slope resistances r_{ce} and $r_{b'c}$ are introduced to describe this effect and may be calculated from

$$r_{ce} = \mu r_e \tag{3.13}$$

and

$$r_{b'c} = \mu \beta r_e \tag{3.14}$$

where μ, the amplification factor, lies between 10^3 and 10^5. It is evident that these resistances are usually very much greater than other circuit resistances ($r_{bb'}$, $r_{b'e}$ and R_C) and for this reason r_{ce} and $r_{b'c}$ can be omitted from the model. However, r_{ce} is important in current sources.

The current gain (β) of a BJT falls with increasing frequency as shown in Fig. 3.3 and this behaviour can be represented in the model by defining β as a complex function of frequency, i.e.

Sparkes (1987), Chapter 4, covers this topic in some detail.

We distinguish between the low-frequency β_o and the frequency-dependent β by using the subscript.

$$\beta = \frac{\beta_o}{1 + j.f/f_\beta} \tag{3.15}$$

where f_β is the frequency at which the magnitude of β has fallen to 0.707 of its low-frequency value (β_o). Another characteristic frequency (f_T) of the device is labelled in Fig. 3.3; f_T is defined as the frequency at which the magnitude of β is

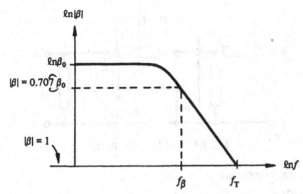

Fig. 3.3   Variation of β with frequency.

unity. The relationship between f_T and f_β is derived from Equation 3.15. For $f \gg f_\beta$,

$$|\beta| = \frac{\beta_0}{f/f_\beta} = 1 \quad \text{at } f = f_T$$

Therefore

$$f_T = \beta_0 f_\beta \tag{3.16}$$

However, it is preferred to retain the physical basis of the model by incorporating the device capacitances which cause the high-frequency deterioration of β, namely the emitter-base diffusion capacitance ($C_{b'e}$) and the transition capacitance ($C_{b'c}$) of the reverse-biased collector-base junction. It can be shown that

$$f_T = \frac{1}{2\pi r_e (C_{b'e} + C_{b'c})} \tag{3.17}$$

and, since manufacturers' data usually include figures for f_T and $C_{b'c}$, the capacitance $C_{b'e}$ can be calculated from Equation 3.17. (Note that $C_{b'e}$ is proportional to I_C and $C_{b'c}$ is proportional to $V_{CB}^{-1/3}$.)

In the hybrid-π model the collector current generator is labelled $g_m v_{b'e}$, where g_m is the mutual conductance, or transconductance, of the BJT and can be calculated as follows.

$$i_c = \beta i_b = \beta \frac{v_{b'e}}{r_{b'e}} = \frac{\beta}{(1+\beta)} \frac{v_{b'e}}{r_e}$$

$$= \frac{\alpha}{r_e} v_{b'e}$$

$$\equiv g_m v_{b'e}$$

where

$$g_m = \frac{\alpha}{r_e} \approx \frac{1}{r_e} \tag{3.18}$$

Ignoring r_{ce} and $r_{b'c}$, since β is defined as the current gain with the output short circuit and $r_{b'c} \gg r_{b'e}$.

49

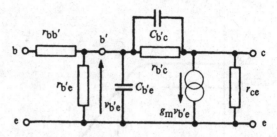

Fig. 3.4   Full hybrid-π model.

Since, at room temperature,

$$r_e \approx \frac{0.025}{I_{E(d.c.)}}\ \Omega$$

(I_E in A)

$$g_m \approx 40 I_{E(d.c.)}\ S \tag{3.19}$$

This voltage-controlled current source ($g_m v_{b'e}$) is equivalent to the current-controlled current source (βi_b) and, since high-frequency behaviour is described by $C_{b'e}$ and $C_{b'c}$, the parameters β and g_m are both real quantities independent of frequency.

The full hybrid-π model (Fig. 3.4) is accurate up to a frequency of approximately $f_T/3$. Table 3.1 summarizes its constituent parameters.

Table 3.1

$r_{bb'}$	typically $100\ \Omega$
$r_{b'e}$	$(1 + \beta)r_e$ where $r_e \approx 0.025/I_E$
r_{ce}	μr_e $\left.\begin{array}{c}\\ \\\end{array}\right\}$ $10^3 \leqslant \mu \leqslant 10^5$, usually omit
$r_{b'c}$	$\mu \beta r_e$
g_m	$\alpha/r_e \approx 1/r_e$
$C_{b'c}$	from data
$C_{b'e}$	calculate from f_T, $C_{b'c}$ and r_e

Simple common-base model

The initial common-emitter model of Fig. 3.1a can be redrawn with its base as the reference terminal common to both input and output. If the directions of the input voltage and the currents are reversed (this is consistent since all directions are changed), the simple common-base model results.

Considering the a.c. equivalent circuit of the biased common-base amplifier of Fig. 3.5, analysis yields

$$\text{Input resistance, } r_{in} = \frac{v_{in}}{i_{in}} = R_E \| r_e \tag{3.20}$$

$$= R_E \| r_{in(t)}$$

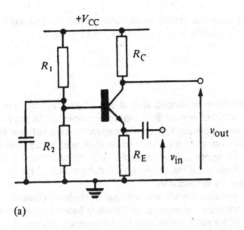

(a)

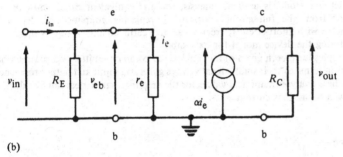

(b)

Fig. 3.5 Common-base amplifier and its a.c. equivalent circuit.

where

$$r_{in(t)} = \frac{v_{eb}}{i_e} = r_e \qquad (3.21)$$

a factor $1/(1 + \beta)$ lower than the common-emitter input resistance.
Also

$$v_{out} = \alpha i_e R_C$$

and

$$v_{in} = i_e r_e$$

Therefore

$$\text{Voltage gain, } A_V = \frac{v_{out}}{v_{in}} = \frac{\alpha R_C}{r_e} \approx \frac{R_C}{r_e} \qquad (3.22)$$

the same magnitude as for the common-emitter amplifier but without inversion
between input and output (no minus sign).

Here the collector current source is considered to be ideal (infinite impedance) and the output resistance of the circuit is simply equal to R_C.

Summary

In this chapter we have developed several small-signal models which are used to analyse the a.c. performance of BJT amplifier circuits. The simplest models for common-emitter and common-base configurations contain only one slope resistance and a collector current generator and, while their use permits rapid analysis, the results can only be approximate and refer only to operation at low frequency. However, the dependence of the results on the current gain of the device and its bias current is immediately predictable.

The hybrid-π equivalent circuit, an extension of the basic common-emitter model, provides a more accurate description by including base resistance and components to describe base-width modulation and high-frequency effects.

Whichever model is used in analysis, the a.c. equivalent circuit must be constructed from the full circuit diagram by replacing coupling and decoupling capacitors with short-circuits, treating as a.c. earth all d.c. supply voltages and substituting the device model for its symbol.

Examples have been given of a.c. analysis of common-emitter and common-base circuits in terms of their small-signal voltage gains and input and output resistances. From these analyses numerical values for the amplifier parameters can be calculated for given β and bias current.

Problems

3.1 Using a simple BJT model, calculate the small-signal voltage gain, input resistance and output resistance for the common-emitter circuit of Problem 2.3. All capacitors may be considered to be short-circuit.

3.2 Repeat Problem 3.1 for the circuit of Problem 2.8. Assume $\beta = 100$.

3.3 Repeat Problem 3.1 for the circuit of Problem 2.11. Assume $\beta = 200$.

3.4 With reference to Fig. 3.6, $V_{CE} = 4.6\,\text{V}$ and $V_{BE} = 0.7\,\text{V}$ under quiescent conditions. Calculate β and r_e for the transistor. Assuming that the input coupling capacitor is sufficiently large for its reactance to be ignored, calculate the mid-band voltage gain (v_o/v_t). Using Equation 2.18, calculate the frequency at which the low frequency voltage gain (v_o/v_t) is 0.707 of its mid-band value.

3.5 Repeat Problem 3.4 assuming an extrinsic base resistance $(r_{bb'})$ of $100\,\Omega$ incorporated in the BJT model.

3.6 Prove that the small-signal voltage gain of a common-emitter amplifier is directly proportional to the supply voltage assuming that the circuit is always biased for maximum output signal handling.

3.7 For the common-base amplifier of Fig. 3.5, calculate the small-signal voltage gain, input resistance and output resistance. $R_1 = 33\,\text{k}\Omega$, $R_2 = 12\,\text{k}\Omega$, $R_C = 2.7\,\text{k}\Omega$, $R_E = 1.5\,\text{k}\Omega$ and $V_{CC} = +15\,\text{V}$. Assume that the capacitors are short-circuit at frequencies of interest.

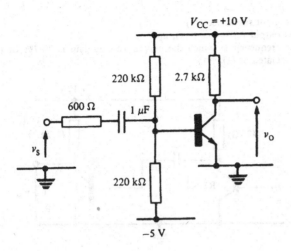

$V_{CC} = +10\,V$

220 kΩ 2.7 kΩ

600 Ω 1 µF

v_s

v_O

220 kΩ

−5 V

Fig. 3.6

3.8 In a potentiometer biased n–p–n common-emitter amplifier, the supply voltage is +6 V and the required peak-to-peak output voltage swing is 4 V symmetrical about the quiescent point. If the quiescent collector current is 1 mA and $V_{CE(sat)} = 0\,V$,

   (a) calculate the value of R_C which allows the quiescent V_E to be maximum and the quiescent collector voltage to be mid-way between its extreme values,

   (b) if $I_C \approx I_E$, calculate the corresponding value of R_E.

   If all capacitors are considered to be short-circuit,

   (c) calculate the small-signal voltage gain,

   (d) if the transistor β is 100, calculate the input resistance assuming that the bias resistors are sufficiently large as to be neglected.

3.9 Measurements on a particular bipolar transistor (biased at $I_C = 2\,mA$, $V_{CE} = 6\,V$) yielded: $f_T = 120\,MHz$, $\beta_o = 100$, $r_{bb'} = 100\,\Omega$, $C_{b'c} = 5\,pF$. Construct, with values, the hybrid-π model for the device working in common-emitter configuration under the above bias conditions.

3.10 Without recourse to using equations, explain why a common-base amplifier is non-inverting.

3.11 A cascade of two common-emitter amplifiers is shown in Fig. 3.7. For simplicity, biasing resistors have been omitted since they have only a small effect on the performance of the circuit. Both transistors have a common-emitter current gain of 100; transistor TR1 is biased at a quiescent collector current of 100 µA and TR2 at 1 mA. Using small-signal a.c. analysis, calculate:

   (a) the overall voltage gain (v_{out}/v_{in}) at frequencies where the reactance of the coupling capacitor ($C = 1\,\mu F$) is negligible;

(b) the input resistance (r_{in});

(c) the output resistance (r_{out}); and

(d) the frequency at which the overall voltage gain is 70.7% of the value calculated in (a).

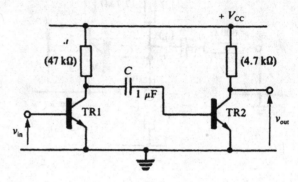

Fig. 3.7

Single-stage BJT amplifiers with feedback   4

Objectives

□ To describe the effects of applying negative feedback to a common-emitter amplifier, namely

improved designability of the transfer function in terms of resistor ratios, and
increase or reduction of r_{in} and r_{out}.

□ To explain why an emitter follower is a good buffer amplifier.
□ To calculate the low-frequency response of a common-emitter amplifier with emitter decoupling.
□ To design amplifier circuits with a high input resistance.

In Chapter 2 negative feedback was applied to the common-emitter amplifier to stabilize the bias conditions of the transistor. In both shunt and potentiometer bias circuits the feedback applied for both a.c. (signal) and d.c. conditions and circuit modifications, e.g. the inclusion of an emitter decoupling capacitor, were introduced to remove the a.c. feedback leaving a stably biased common-emitter amplifier. The question arises: In what way would the signal behaviour of the amplifier be modified if the a.c. feedback was retained?

Series feedback amplifier

If the emitter decoupling capacitor is removed from the common-emitter circuit, Fig. 4.1 results; the a.c. feedback now present in the circuit is called **series** negative feedback since the input voltage is backed-off by an i_e-dependent voltage in series with the controlling v_{be}.

We have seen in Chapter 1 that V_{BE} remains approximately constant with changes in device current: a factor of 10 times change in current needs only a 60 mV change in V_{BE}! Hence, the input signal (v_{in}) at the base is almost exactly reproduced at the emitter causing an emitter current change

$$i_e \approx \frac{v_{in}}{R_E} \tag{4.1}$$

Since the output current $i_c \approx i_e$,

$$i_c \approx \frac{v_{in}}{R_E}$$

or

$$\frac{i_c}{v_{in}} \approx \frac{1}{R_E} \tag{4.2}$$

Assuming that $v_{be} = 0$. This is an a.c. voltage *not* to be confused with the d.c. V_{BE} (≈ 0.7 V).

Upper case symbols are used for d.c. quantities, lower case for a.c.

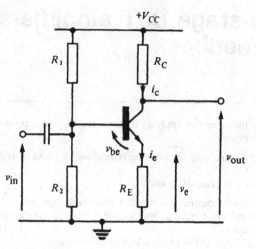

Fig. 4.1 Series feedback amplifier.

The series feedback amplifier may be regarded fundamentally as a circuit with a transconductance ($g_T = 1/R_E$) specified by a resistor value unlike the common-emitter amplifier whose transconductance ($g_m = 1/r_e$) is both bias and temperature dependent, as well as being nonlinear.

If the collector current is converted to a voltage as

$$v_{out} = -i_c R_C$$

then

Simply the ratio of two resistors and independent of bias and β.

$$\text{Voltage gain, } A_V = \frac{v_{out}}{v_{in}} \approx -\frac{R_C}{R_E} \tag{4.3}$$

This approximate relationship is called a simple rule of thumb, which is very useful for design purposes. However, analysis of the a.c. equivalent circuit must be performed to provide accurate expressions for the voltage gain and for the input and output resistances.

With reference to the a.c. equivalent circuit of the series feedback stage (Fig. 4.2),

$$\boxed{v_{in} = i_b r_{be} + (i_b + \beta i_b)R_E}$$

$$= i_b(r_{be} + (1 + \beta)R_E)$$

Therefore

This approach to the analysis involves us in systematically generating results as we proceed. As an alternative method of solution, we can extract from the equivalent circuit the three independent equations (shown blocked). Since we have three equations in four variables, any pair of variables now may be related to each other as $r_{in(t)}$, g_T and A_V.

$$r_{in(t)} = \frac{v_{in}}{i_b} = r_{be} + (1 + \beta)R_E \tag{4.4}$$

and

$$r_{in} = r_{in(t)} \| R_1 \| R_2 \tag{4.5}$$

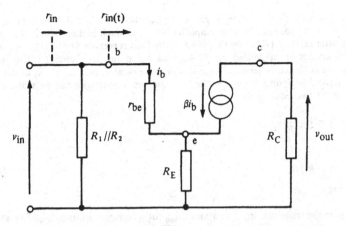

Fig. 4.2 An a.c. equivalent circuit of series feedback amplifier.

Now

$$i_c = \beta i_b$$

$$= \beta \frac{v_{in}}{r_{in(t)}}$$

$$= \frac{\beta v_{in}}{r_{be} + (1 + \beta)R_E}$$

Therefore

$$g_T = \frac{i_c}{v_{in}} = \frac{\beta}{r_{be} + (1 + \beta)R_E}$$

$$\approx \frac{1}{r_e + R_E} \quad \text{if} \quad \beta \gg 1 \tag{4.6}$$

and

$$v_{out} = -i_c R_C = -g_T v_{in} R_C$$

Therefore

$$A_V = \frac{v_{out}}{v_{in}} = -\frac{R_C}{r_e + R_E} \tag{4.7}$$

Equations 4.6 and 4.7 differ from the simple expressions of Equations 4.2 and 4.3 by the inclusion of the r_e term which describes the action of the emitter-base junction. In practice this term may be ignored since R_E usually has a value of several kΩ while r_e is of the order of tens of Ω, at most, for collector currents in the mA range.

Using this alternative method, it is essential to find a number of independent equations which is one less than the number of variables. Otherwise, we can be most frustrated by curious results (such as 1 = 1) appearing in our solution attempt!

The degree of nonlinearity is determined by the significance r_e.

The transistor input resistance, given by Equation 4.4, is still β-dependent as was the case in the common-emitter amplifier but is of much higher value. (Note that the emitter resistor (R_E) when referred to the base is seen as $(1 + \beta)R_E$, just as r_e in the emitter is magnified by $(1 + \beta)$ to r_{be} in the base circuit.) The input resistance at the input terminal (r_{in}) is now more closely specified by R_1 in parallel with R_2 and designing for a specific bias and input resistance can be realized by solving the two equations

$$V_B = V_{CC} \frac{R_2}{R_1 + R_2} \tag{4.8}$$

and

$$r_{in} = R_1 \| R_2 \tag{4.9}$$

ignoring $r_{in(t)}$.

Exercise 4.1   Compare the transistor input resistance $r_{in(t)}$ for a common-emitter amplifier and a series feedback stage. In both cases the BJT is biased at $I_C = 1$ mA and you may assume that $\beta = 100$ and $R_E = 2.2$ kΩ.
[*Answer*: 2.5 kΩ, 220 kΩ, approximately.]

Observation of the a.c. equivalent circuit shows that the series feedback amplifier output resistance is simply R_C (in parallel with the infinite output resistance of the ideal collector current source), the same as for the common-emitter circuit. This, however, does not take into account the non-infinite transistor output resistance represented by r_{ce} and $r_{b'c}$ in the full hybrid-π model. If this full model is used in analysis the output resistance of the common-emitter stage is approximately equal to r_{ce} and, for the series feedback amplifier, can be shown to be (neglecting $r_{b'c}$):

$$r_{out(t)} = r_{ce}\left(1 + \frac{\beta R_E}{r_{be} + R_E}\right) + r_{be}\| R_E \tag{4.10}$$

a value which is significantly higher than for the common-emitter case. This is significant when the biased transistor is to be used as a current source without a collector load resistor; leaving the emitter undecoupled yields a much higher output resistance (and closer to the ideal current source) than that of the common-emitter circuit. When R_C is included to convert the circuit into a voltage amplifier, the contribution of the transistor output resistance to the circuit output resistance ($\approx R_C$) is usually minimal in both cases.

Series feedback has increased both input and output resistances compared with the common-emitter circuit. The voltage gain, although reduced, is now accurately defined by a resistor ratio.

Exercise 4.2   Estimate the transistor output resistance (Equation 4.10) for a series feedback stage assuming $I_C = 1$ mA, $\beta = 100$, $R_E = 2.2$ kΩ and $\mu = 5 \times 10^3$. ($r_{ce} = \mu r_e$)
[*Answer*: approx. 6 MΩ, i.e. $\gg r_{ce}$.]

Design Note: In the basic series feedback circuit of Fig. 4.1, the emitter resistor (R_E) performs two functions: d.c. negative feedback for stable biasing and a.c. negative feedback for signal transconductance and voltage gain specification. In a circuit with a prescribed collector load (R_C) it may appear impossible to calculate a value for R_E which provides, for bias, a d.c. emitter voltage of some 2 to 3 V and at the same time allows the voltage gain specification to be met. For example, with $I_C = 1$ mA, an R_E of 2.2 kΩ provides a satisfactory d.c. emitter voltage. If a

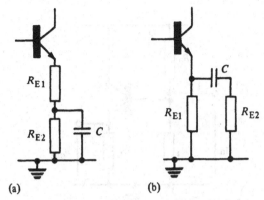

(a)   (b)

Fig. 4.3   Modifications to emitter circuit.

voltage gain of -10 and a 10 kΩ collector load are specified, then, since $|A_v| \approx R_C/R_E$, R_E would have to be reduced to 1 kΩ, rather low for biasing purposes.

This apparent conflict does not have to be solved by compromise. Noting that the d.c. value of R_E should be 2.2 kΩ and that its a.c. value should be 1 kΩ, the emitter circuit should be modified to one of the configurations shown in Fig. 4.3. In Fig. 4.3a,

$$R_{E(d.c)} = R_{E1} + R_{E2}$$

and   (4.11)

$$R_{E(a.c)} = R_{E1}$$

while for Fig. 4.3b

$$R_{E(d.c.)} = R_{E1}$$

and   (4.12)

$$R_{E(a.c.)} = R_{E1}\|R_{E2}$$

In both cases the capacitor (C) is assumed to be sufficiently large as to be considered an a.c. short-circuit at the signal frequency.

This can now be solved by using $R_{E1} = 1$ kΩ and $R_{E2} = 1.2$ kΩ in Fig. 4.3a or $R_{E1} = 2.2$ kΩ and $R_{E2} = 1.8$ kΩ in Fig. 4.3b.

The emitter follower

The emitter follower or common-collector amplifier (Fig. 4.4) is best considered as a series feedback amplifier in which the output is taken from the emitter rather than from the collector.

In simple terms, the output voltage is separated from the input voltage by the almost constant voltage drop across the forward biased base-emitter junction. As the input voltage varies, so does the output in almost exact sympathy except for the 0.7 V d.c. offset. The emitter is said to follow the base, hence the term **emitter follower**,

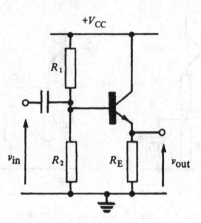

Fig. 4.4 Emitter follower.

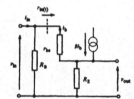

and the circuit is expected to have a voltage gain of unity (+1 since there is no signal inversion between base and emitter).

Analysis of the a.c. equivalent circuit is necessary to yield accurate expressions for voltage gain, input resistance and output resistance.

The procedure for determining the input resistance to the transistor $r_{in(t)}$ is identical to that for the series feedback amplifier and yields the same result, i.e.

$$r_{in(t)} = r_{be} + (1 + \beta)R_E \qquad (4.13)$$

Now

$$v_{out} = (i_b + \beta i_b)R_E = i_b(1 + \beta)R_E$$

and

$$v_{in} = i_b r_{be} + (i_b + \beta i_b)R_E$$
$$= i_b[r_{be} + (1 + \beta)R_E]$$

Therefore

$$A_V = \frac{v_{out}}{v_{in}} = \frac{(1 + \beta)R_E}{r_{be} + (1 + \beta)R_E} \simeq \frac{R_E}{r_e + R_E} \qquad (4.14)$$

Equation 4.14 shows that the voltage gain is always less than unity but is very close to unity if R_E is much greater than r_e, which is usually the case.

The output resistance of the emitter follower is derived by the method outlined in Chapter 3, i.e. applying a voltage signal to the output and calculating the resultant current flow. The input voltage source is replaced by a short-circuit. It should be noted that R_E is directly across the output and, to ease analysis, can be removed and reintroduced afterwards as a parallel component.

$$v_o = (i_o + \beta i_b)r_{be}$$

and

$$i_b = -(i_o + \beta i_b)$$

60

Therefore

$$i_b = -\frac{i_o}{1 + \beta}$$

Substituting gives

$$v_o = i_o\left(1 - \frac{\beta}{1 + \beta}\right)r_{be}$$

$$= i_o\frac{r_{be}}{1 + \beta} = i_o r_e$$

Therefore

$$\frac{v_o}{i_o} = r_e$$

and, reintroducing R_E,

$$r_{out} = r_e \| R_E \tag{4.15}$$

$$\approx r_e$$

since $r_e \ll R_E$.

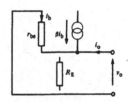

Equation 4.15 applies to an emitter follower driven by an ideal voltage source.

Exercise 4.3

Prove that the output resistance of an emitter follower when driven by a generator of internal resistance (R_S) is given by

$$r_{out} = \left(r_e + \frac{R_S}{1 + \beta}\right)\| R_E$$

(Note that this expression reduces to Equation 4.15 if $R_S = 0$; also, that resistance in the base circuit, when referred to the emitter circuit, is divided by $(1 + \beta)$.)

Prove that, if R_S is non-zero, the voltage gain of an emitter follower is reduced to

$$A_V = \frac{R_E}{R_E + r_e + \dfrac{R_S}{1 + \beta}}$$

Also, prove that, for a series feedback stage driven from a non-zero source resistance, the voltage gain is reduced by an additional term to

$$A_V = \frac{-R_C}{R_E + r_e + \dfrac{R_S}{1 + \beta}}$$

We now have a detailed knowledge of the properties of an emitter follower: voltage gain very close to unity, very high input resistance and very low output resistance.

These are the very desirable properties of a buffer amplifier which serve to reduce loading effects on signal transmission between a source and load.

Consider the circuit of Fig. 4.5a, which is not buffered.

$$v_L = v_s\frac{R_L}{R_S + R_L}$$

61

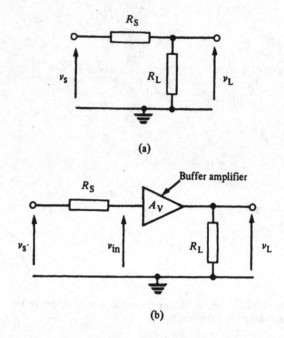

(a)

(b)

Fig. 4.5 Use of buffer amplifier.

However, if an ideal buffer amplifier ($r_{in} = \infty$, $r_{out} = 0$, $A_V = 1$) is introduced as in Fig. 4.5b, then

$$v_{in} = v_s \text{ and } v_L = A_V v_{in}$$

giving

$$v_L = v_s$$

i.e. unattenuated transmission between source and load.

Exercise 4.4 Calculate the output resistance of an emitter follower biased at $I_C = 5$ mA and with an emitter resistor (R_E) of 1 kΩ. Also, calculate the voltage gain of the unloaded circuit. In both cases assume $R_S = 0$.
If an external load ($R_L = 100$ Ω) is capacitively coupled to the output, recalculate the voltage gain.
[*Answer*: $r_{out} \approx 5$ Ω, $A_V \approx 0.995$; $A_V \approx 0.948$.]

Effect of emitter decoupling capacitor

It has been explained that, to turn the potentiometer bias circuit into a common-emitter amplifier, the emitter resistor (R_E) is bypassed with a high-valued capacitor

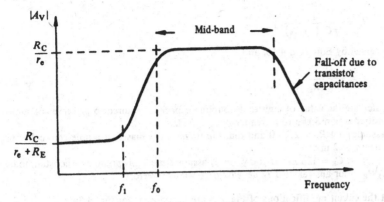

Fig. 4.6 Low-frequency response of common-emitter amplifier caused by the decoupling capacitor.

(C_E). This capacitor cannot act as a short-circuit for all signal frequencies down to d.c.; in fact, at d.c., C_E is an open circuit and the amplifier behaves as a series feedback circuit with a voltage gain equal to $-R_C/(r_e + R_E)$. This is usually significantly less than the voltage gain of the common-emitter amplifier ($-R_C/r_e$) realized at frequencies where the reactance of C_E is very small. The resultant frequency response, modulus of voltage gain plotted against frequency, is shown in Fig. 4.6. The critical frequency (f_o), at which the voltage gain is 0.707 of its midband level, may be calculated in one of two ways.

f_o is given by

$$f_o = \frac{1}{2\pi C_E (\text{Resistance at emitter})}$$

and, since the resistance at the emitter is the output resistance of the emitter follower ($= r_e \| R_E$),

$$f_o = \frac{1}{2\pi C_E (r_e \| R_E)} \approx \frac{1}{2\pi C_E r_e} \tag{4.16}$$

Alternatively, the circuit may be analysed without assuming that C_E is an a.c. short-circuit. The resultant expression for the voltage gain should be converted into the standard form

The reader should perform this analysis.

$$A_V = [\text{d.c. voltage gain}] \frac{[1 + \mathrm{j}f/f_1]}{[1 + \mathrm{j}f/f_o]}$$

where f_o is the **pole** frequency described above, and f_1 is the **zero** frequency, i.e. the characteristic frequency at which the voltage gain is 1.414 times the d.c. voltage gain. Analysis gives

Millman and Grabel (1987), Chapter 11 discuss frequency response in terms of poles and zeros.

$$A_V = \frac{-R_C}{r_e + R_E} \frac{[1 + \mathrm{j}\omega C_E R_E]}{[1 + \mathrm{j}\omega C_E (r_e \| R_E)]} \tag{4.17}$$

from which it is evident that

63

$$f_o = \frac{1}{2\pi C_E (r_e \| R_E)}$$

as given by Equation 4.16, and

$$f_1 = \frac{1}{2\pi C_E R_E} \tag{4.18}$$

Exercise 4.5 Calculate the value of emitter decoupling capacitor required to preserve common-emitter action down to a frequency of 100 Hz.

Assume that $R_E = 2.2$ kΩ and that the transistor is biased at a quiescent collector current of 2 mA.

[*Answer*: $C_E = 128$ μF. If you wrongly assumed the appropriate time constant to be $C_E R_E$, your answer of 0.72 μF would be wildly in error!]

If the circuit modifications of Fig. 4.3 are introduced, for Fig. 4.3a,

$$f_o = \frac{1}{2\pi C [R_{E2} \| (R_{E1} + r_e)]} \tag{4.19}$$

and for Fig. 4.3b,

$$f_o = \frac{1}{2\pi C (R_{E2} + R_{E1} \| r_e)} \tag{4.20}$$

Shunt feedback amplifier

In the initial stages of circuit design we require simple equations describing the approximate circuit performance to produce the component values. Subsequent application of the more precise relationships can be used to verify the design.

Shunt negative feedback has already been considered in Chapter 2 with regard to providing stable d.c. bias for a transistor. Normally, to calculate the a.c. performance of the circuit (Fig. 4.7), one would analyse its a.c. equivalent circuit but, in this case, the results of accurate analysis are cumbersome and of little use in the design process. Here we will consider an approximate analysis and only quote, without derivation, the more detailed results.

If a voltage signal source is applied directly to the input of a shunt feedback amplifier (i.e. at the transistor base), the feedback current through the feedback resistor (R_B) is shunted to earth through the zero resistance of the voltage source rather than act on the base. The amplifier then behaves as a common-emitter stage with a collector load of R_C in parallel with R_B and no feedback results. Therefore, a shunt feedback amplifier must be driven from a non-zero source impedance to allow the feedback to operate.

To show how the circuit operates let us make two assumptions. Assume, first, that the BJT has an infinite β making the signal base current (i_b) equal to zero and, second, that the base-emitter voltage of the transistor does not change with the applied signal (i.e. $v_{be} = 0$).

Input source current (i_{in}) now flows through the feedback resistor creating a voltage drop across it equal to ($-i_{in} R_B$) and, since $v_{be} = 0$,

$$v_{out} = -i_{in} R_B$$

or the transresistance,

$$r_T = \frac{v_{out}}{i_{in}} = -R_B \tag{4.21}$$

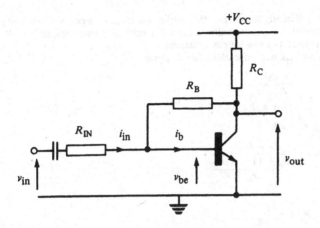

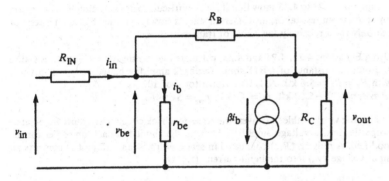

Fig. 4.7 Basic shunt feedback amplifier with its a.c. equivalent circuit.

Since v_{be} is assumed to be zero, the input resistance of the circuit is also zero or at least, in practice, very low.

To estimate the output resistance, apply a voltage (v_o) at the output and calculate the resultant current (i_o) with the input open-circuit (since the amplifier is driven from a current source). (Note that R_C is in parallel with the output and therefore can be removed for analysis, reincorporating it later.) Now

$$i_o = i_c + i_b = (1 + \beta)i_b$$

and

$$v_o = i_b R_B (\text{if } v_{be} = 0)$$

Therefore

$$r_{out} = \frac{v_o}{i_o} = \frac{R_B}{1 + \beta} \| R_C \tag{4.22}$$

The input resistance of a circuit is always reduced when shunt negative feedback is applied.

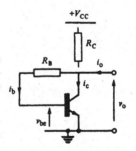

65

a figure significantly lower than the output resistance of a common-emitter stage.

To summarize, a shunt feedback amplifier exhibits a transresistance of R_B, low input resistance and low output resistance.

Analysis of the a.c. equivalent circuit yields

$$r_T = \frac{-R_B}{1 + \dfrac{R_B + r_{be}}{(1 + \beta)R_C}} \tag{4.23}$$

(assuming $R_B \gg r_e$).

The reader can gain useful practice in analysis by justifying these results.

$$r_{in} = \frac{R_C + R_B}{1 + \dfrac{R_B + (1 + \beta)R_C}{r_{be}}} \tag{4.24}$$

and

$$r_{out} = \frac{R_B + r_{be}}{1 + \beta} \, \| R_C \tag{4.25}$$

Equations 4.23 to 4.25 show that it is very difficult to design a circuit for a certain input or output resistance; the transresistance is closely specified by a circuit resistor but only for a relatively low value of transresistance.

Exercise 4.6   Using Equations 4.23, 4.24 and 4.25, calculate the transresistance, input resistance and output resistance of a BJT shunt feedback amplifier biased at $I_C = 1$ mA and with $R_B = R_C = 5.6$ kΩ. Assume a transistor β of 100.
[*Answer*: $r_T = -5.52$ kΩ, $r_{in} = 49.3$ Ω, $r_{out} = 54.9$ Ω.]

Just as it was possible to convert the series feedback circuit with transconductance properties into a voltage amplifier, likewise the shunt feedback amplifier can be modified. A resistor (R_{IN}) connected in series with the input (Fig. 4.7) converts an input voltage (v_{in}) into the input current (i_{in}) as

$$i_{in} = \frac{v_{in}}{R_{IN}}$$

and, since

$$\frac{v_{out}}{i_{in}} \approx -R_B$$

the voltage gain of the resulting circuit is given by

$$A_V = \frac{v_{out}}{v_{in}} \approx -\frac{R_B}{R_{IN}} \tag{4.26}$$

This relationship assumes that the transresistance is given by $-R_B$ and that the input resistance at the base of the transistor is zero. Taking the circuit of Exercise 4.5 and including an input resistor (R_{IN}) of 1 kΩ, the approximate relationships give

$$A_V \approx -5.6$$

However, calculation has shown that the transresistance is actually -5.52 kΩ and, since the input resistance at the transistor base is 49.3 Ω,

$$i_{in} = \frac{v_{in}}{R_{IN} + r_{in}}$$

Therefore

$$A_V = \frac{-r_T}{R_{IN} + r_{in}}$$

$$= -\frac{5.52}{1.0493} = -5.26$$

The precision of the single-stage BJT shunt feedback amplifier is poor in comparison with its operational amplifier counterpart. This is due to the relatively low voltage gain of the common-emitter amplifier to which the feedback is applied. See: Horrocks (1990). However, these calculations justify the use of approximate relationships in design.

showing a discrepancy of some 6% when compared with the ideal relationship of Equation 4.26.

The input resistance of this configuration is $(R_{IN} + r_{in})$, which can be approximated to R_{IN}.

Design a shunt feedback amplifier to meet the approximate specification: $A_V = -10$, $r_{in} = 1$ kΩ. Assume a transistor biased at $I_C = 2$ mA and with $\beta = 100$. The supply voltage is +9 V.
Determine the accurate performance of your circuit.
[*Answer*: $R_C = 2.2$ kΩ, $R_{IN} = 1$ kΩ and $R_B = 10$ kΩ.
$r_T = 9.518$ kΩ, $r_{in} = 1.066$ kΩ, $A_V = -8.93$.]

Exercise 4.7

Design Note: In a shunt feedback amplifier, the feedback resistor can perform two functions: providing d.c. base current for biasing and defining the a.c. transresistance. However, it is seldom that the two functions can be merged into a single resistor; invariably the value of R_B for d.c. biasing is much higher than the value required for the a.c. transresistance.

Using the circuit shown, it is a simple matter to achieve the required result. Here,

$$R_{B(d.c.)} = R_1$$

and

$$R_{B(a.c.)} = R_1 \| R_2$$

assuming that the capacitor (C) has negligible reactance at signal frequencies compared with R_2.

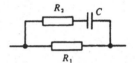

Perform a complete design for the circuit specified in Exercise 4.7.

Design Example 4.1

Solution. Since $V_{CC} = +9$ V, $I_C R_C$ should be approximately $V_{CC}/2$, or 4.5 V, for maximum output signal handling.

With $I_C = 2$ mA, $R_C = 2.2$ kΩ (nearest preferred value), giving $V_C = (V_{CC} - I_C R_C) = 4.6$ V.

To calculate $R_{B(d.c.)}$,

$$R_1 = \frac{V_C - V_{BE}}{I_B}$$

$$= (4.6 - 0.7)/\frac{2 \times 10^{-3}}{100}$$

$$= 195 \text{ k}\Omega$$

The d.c. design is performed first. Modifications to meet the a.c. specification are introduced later.

Take $R_1 = 180$ kΩ (nearest preferred value).
  To calculate R_2,

$$R_{B(a.c.)} = R_1 \| R_2 = 10 \text{ k}\Omega$$

We could select R_2 as 10 kΩ which, in parallel with R_1, would give $R_{B(a.c.)} = 9.47$ kΩ. Alternatively, with $R_2 = 12$ kΩ, $R_{B(a.c.)} = 11.25$ kΩ. This latter value for R_2 is a better solution since, with $R_{B(a.c.)} = 11.25$ kΩ,

$$r_T = -10.65 \text{ k}\Omega \quad \text{and} \quad r_{in} = 72.35 \text{ }\Omega$$

and, with $R_{IN} = 1$ kΩ, results in a voltage gain of -9.93.

High-input resistance techniques

We have seen that a voltage amplifier should exhibit a very high input resistance in order to avoid signal attenuation between the circuit input resistance and the source resistance of the input generator. The two circuits already encountered which have an inherently high input resistance are the series feedback amplifier and the emitter follower, each with an input resistance looking into the transistor base ($r_{in(t)}$) given by

$$r_{in(t)} = r_{be} + (1 + \beta)R_E$$

Recognizing that $r_{in(t)}$ is directly proportional to β, it is evident that, for a maximum value of input resistance, the transistor type should be selected for maximum β. Even with modern transistors, β is seldom guaranteed to be greater than 200 and the technique of using two BJTs in a compound connection to give a very high current gain can prove very useful.

Compound-connected BJTs

Consider the two n–p–n BJTs connected as shown in Fig. 4.8 to form a composite n–p–n transistor. The base current of TR2 is the emitter current of TR1 and is magnified by TR2 to yield, for the compound pair

$$I_E = (1 + \beta_1)(1 + \beta_2)I_{B1} \approx \beta_1\beta_2 I_{B1} \tag{4.27}$$

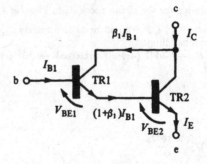

Fig. 4.8 Composite n–p–n connection.

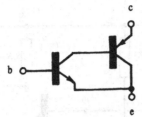

Fig. 4.9 Improved compound connection (n-p-n).

and

$$I_C = \beta_1 I_{B1} + \beta_2 (1 + \beta_1) I_{B1} \approx \beta_1 \beta_2 I_{B1} \qquad (4.28)$$

where I_{B1} is the base current of transistor TR1 and β_1, β_2 are the common-emitter current gains of TR1 and TR2 respectively. This compound connection, sometimes known as a **Darlington pair**, behaves as a single transistor with a current gain $\beta \approx \beta_1 \beta_2$ but with a base-emitter voltage of $V_{BE1} + V_{BE2}$, or approximately 1.4 V, which may be embarrassingly high for circuits with a very low supply voltage.

Darlington pairs are used in examples of power amplifiers and power supply regulator designs (Chapters 8 and 9).

A compound pair with an effective V_{BE} of approximately 0.7 V can be realized by combining an n-p-n transistor (TR1) with a p-n-p transistor (TR2) as shown in Fig. 4.9.

This connection also has an overall current gain given by the product of the current gains of the individual transistors.

Compound p-n-p transistors with a high β are produced by interchanging the transistor types in Figs 4.8 and 4.9. This technique is widely used in integrated circuit design to enhance the β of integrated p-n-p transistors which are difficult to fabricate with a high β in an essentially n-p-n process.

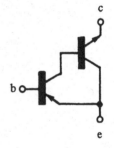

Bootstrap bias circuit

A major problem when designing a circuit for high input resistance is the shunting effect of the base bias resistors, $R_B (= R_1 \| R_2$ for a potentiometer biased circuit). If this effect is isolated, a truly high circuit input resistance would result. The technique used is called **bootstrapping** and involves adding an extra resistor to the bias circuit and magnifying its effective a.c. value by a capacitive coupling which aims to produce identical signal voltages at both ends of the resistor. Then, ideally, no signal current is taken by the bias circuit and the input resistance is defined by the input resistance to the transistor.

This is a special application of Miller's theorem, see Millman and Grabel (1987), p. 849.

In the circuit of Fig. 4.10, bootstrapping has been incorporated in the bias circuit of a series feedback amplifier. For d.c. bias, the transistor base has a Thévenin source resistance of $(R_3 + R_1 \| R_2)$ which, as described in Chapter 2, must be kept low to minimize the variability of the base voltage due to β spread.

With the addition of bootstrapping components (R_3 and C) and assuming that C is of negligible reactance at signal frequencies, the a.c. value of the emitter resistance is given by

$$R_E' = R_E \| R_1 \| R_2 \qquad (4.29)$$

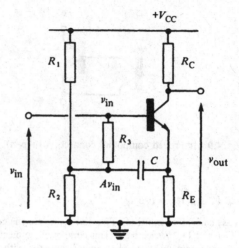

Fig. 4.10 Bootstrap bias circuit.

In practice this represents a small reduction in R_E.

Now, the voltage gain of an emitter follower with emitter resistance R_E' is

$$A = \frac{R_E'}{r_e + R_E'} \tag{4.30}$$

which is very close to unity. Hence, with an input signal (v_{in}) applied to the base, the signal which appears at the emitter ($A v_{in}$) is applied to the lower end of R_3. Therefore, the signal voltage appearing across R_3 is $(1 - A)v_{in}$, very much less than the full input signal, and R_3 now appears to have an effective value (for a.c. signals) of

$$R_3' = \frac{R_3}{1 - A} \gg R_3 \tag{4.31}$$

The input resistance of the complete circuit is

$$r_{in} \approx [r_{be} + (1 + \beta)R_E'] \| R_3' \tag{4.32}$$

compared with the figure of $[r_{be} + (1 + \beta)R_E] \| R_1 \| R_2$ for the unmodified circuit.

The voltage gain of the circuit is given by

$$A_V \approx -\frac{R_C}{R_E'} \tag{4.33}$$

The expression for input resistance given by Equation 4.32 is approximate since sections of the circuit have been considered in isolation and then combined without full regard for interaction of the sections. Analysis of the a.c. equivalent circuit yields the accurate expression for the input resistance of the bootstrap circuit

The reader should perform this analysis for further practice.

$$r_{in} = R_3 \| r_{be} + \left(1 + \frac{\beta R_3}{r_{be} + R_3}\right) R_E' \tag{4.34}$$

70

This relationship is more difficult to use than that of Equation 4.32 which, though approximate, does show distinctly the two separate contributions made to the overall input resistance by the transistor and by the bootstrapped bias circuit.

Using the circuit of Fig. 4.10, design a series feedback amplifier to provide a voltage gain of −2 and an input resistance of 100 kΩ. Assume $V_{CC} = +12$ V and $\beta \geqslant 100$.

Solution. The initial step in the design is to apportion the total circuit input resistance between the transistor and the bias circuit. A sensible first attempt would assume that these two components are equal, each double the total input resistance. If it is found that one of these is difficult to realize, the partitioning may be altered on a second iteration to relieve the problem. With this assumption, Equation 4.32 gives

$$r_{in(t)} = r_{be} + (1 + \beta)R'_E \geqslant 200 \text{ k}\Omega$$

and

$$R'_3 + R'_E \geqslant 200 \text{ k}\Omega$$

Using the dominant terms to give simpler equations,

$$\beta R'_E \geqslant 200 \text{ k}\Omega \quad \text{and} \quad R'_3 \geqslant 200 \text{ k}\Omega$$

With a minimum β of 100, $R'_E \geqslant 2$ kΩ and R_E is chosen as 2.7 kΩ to allow for the shunting effect of R_1 and R_2.

To meet the voltage gain specification, R_C is approximately 4 kΩ (calculated later when R_1 and R_2 are known).

Since the supply voltage is +12 V, a suitable bias current for the transistor is 1 mA thus allowing approximately half the supply for quiescent V_{CE} and achieving a good output voltage swing capability.

Bias components R_1 and R_2 can now be calculated.

$$V_B = I_E R_E + V_{BE} = 2.7 + 0.7 = 3.4 \text{ V}$$

and

$$I_{B(max)} = I_C / \beta_{(min)} = 10 \text{ } \mu\text{A}$$

With R_2 chosen as 33 kΩ, the bias chain current (I) through R_2 is 103 μA and

$$R_1 = \frac{V_{CC} - V_B}{I + I_B} = 76 \text{ k}\Omega$$

or 82 kΩ (nearest preferred value). Therefore

$$R'_E = R_1 \| R_2 \| R_E = 2.42 \text{ k}\Omega$$

The emitter follower voltage gain (A) is calculated as

$$A = \frac{R'_E}{r_e + R'_E} = 0.99$$

and since

$$R'_3 = \frac{R_3}{1 - A} \geqslant 200 \text{ k}\Omega$$

Design Example 4.2

In circuit design, compromises and sensible assumptions usually have to be made.

Here we apply the bias design procedure presented in Chapter 2.

R_3 should be selected as 2.2 kΩ.

To establish a voltage gain of −2 for the series feedback circuit,

$$R_C = 2 \times R_E' = 5.6 \text{ k}\Omega$$

(nearest preferred value). The design is now complete

$$R_C = 5.6 \text{ k}\Omega$$
$$R_E = 2.7 \text{ k}\Omega$$
$$R_1 = 82 \text{ k}\Omega$$
$$R_2 = 33 \text{ k}\Omega$$
$$R_3 = 2.2 \text{ k}\Omega$$
$$C = \text{large}$$

Try substituting values into the approximate Equation 4.32. This should prove the value of the piece-wise consideration.

and may be checked against the input resistance specification. Substituting values into Equation 4.34 gives $r_{in} = 116$ kΩ, which is satisfactory.

Summary

This chapter has been concerned predominantly with the application of a.c. negative feedback to a common-emitter amplifier and determining the performance of these circuits by analysis of their a.c. models.

In the series and shunt amplifiers we have discovered that their transfer performance, transconductance and transresistance, respectively, are determined to a close approximation by a circuit resistor value with only a secondary contribution from transistor parameters. However, input and output resistances remain very dependent on device parameters but have been significantly raised (series) or lowered (shunt) by the negative feedback. As a result, the designability of single-stage BJT amplifiers has been improved to an extent which cannot be matched by the use of a common-emitter amplifier.

The performance of the common-collector amplifier, or emitter follower, has been assessed and its suitability for use as a unity-gain buffer amplifier recognized. In this context, the quest for high input resistance led to discussion of compound-connected pairs of BJTs and the bootstrap bias circuit which significantly reduces the shunting effect of the bias resistor chain.

With an appreciation of the emitter follower and series feedback amplifier it was relatively easy to determine the effect on the low-frequency voltage gain of a common-emitter amplifier caused by the emitter decoupling capacitor.

In treating each amplifier configuration it was considered important to describe the basic mode of operation of the circuits and thus derive simple rule-of-thumb expressions for their performance, later justifying these results by full analysis.

The design process is complex. First, it relies very much on the experience of the designer to establish a circuit configuration which should meet the specification. Second, the results of circuit analysis must be available in a simple form which, with step-by-step application, produce the circuit component values. Next, the design should be checked by substituting calculated values into accurate performance equations and then a prototype can be constructed and tested for final verification. Even with complex circuits, where the simulation is carried out by computer, the essence of the design process is unchanged.

SPICE is a popular simulation program. See Morant (1990), Chapter 9.

Analysis and design go hand in hand. No apology is made for the extensive analytic treatment presented in this chapter; practice in analysis and engendering familiarity with design procedures are essential facets of the training of an electronic circuit designer.

Problems

4.1 Calculate the small-signal a.c. voltage gain and input resistance for the series feedback circuit which results when the emitter decoupling capacitor is disconnected in the circuit of Problem 2.8. Assume $\beta = 100$.

4.2 Repeat Problem 4.1 for the circuit of Problem 2.11 with the emitter decoupling capacitor removed. Assume $\beta = 50$.

4.3 Calculate the small-signal voltage gain, input resistance and output resistance for the circuit of Problem 2.5. Assume $\beta = 100$. If the emitter is now decoupled with a 10 μF capacitor, sketch the behaviour of the low-frequency voltage gain against frequency. (Equation 4.18 will prove useful.)

4.4 In the emitter follower circuit of Fig. 4.4, $V_{CC} = +20$ V, $R_1 = R_2 = 10$ kΩ and $R_E = 2$ kΩ. If the BJT has a β of 200, calculate
(a) the input resistance at the base of the transistor ($r_{in(t)}$),
(b) the input resistance of the complete circuit, and
(c) the voltage gain and output resistance.

4.5 If the circuit of Problem 4.4 is driven from a source resistance of 1 kΩ, determine the voltage gain and output resistance.

4.6 For the circuit of Fig. 4.11, estimate

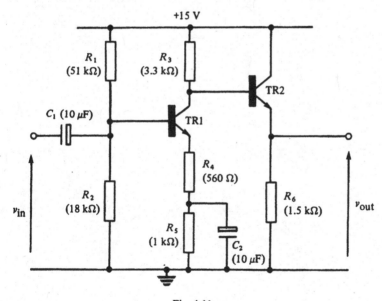

Fig. 4.11

(a) the small-signal voltage gain at frequencies where the capacitors may be considered short-circuit,

(b) the quiescent d.c. output voltage, and

(c) the mid-band input and output resistances of the circuit.

Also, sketch the low-frequency response due to capacitors C_1 and C_2. For both transistors assume that $\beta = 100$ and $V_{BE} = 0.7$ V.

4.7 If the shunt feedback amplifier of Fig. 4.7 is voltage driven at the base of the transistor (i.e. $R_{IN} = 0$), prove that the circuit has a small-signal a.c. voltage gain identical to that of a common-emitter amplifier with collector load $R_C \| R_B$. (Assume $\beta \gg 1$ and $R_B \gg r_e$.)

4.8 Show that the β of the n-p-n/p-n-p compound connection of Fig. 4.9 is $\beta_1(1 + \beta_2)$, where β_1 and β_2 are the βs of the n-p-n and p-n-p transistors respectively.

4.9 Two n-p-n transistors are connected as a Darlington pair (Fig. 4.8). If each transistor is represented by a simple r_{be} plus βi_b model, determine the composite model.

4.10 In the bootstrap bias circuit of Fig. 4.10, $R_C = 1.5$ kΩ, $R_E = 1$ kΩ, $R_1 = 8.2$ kΩ, $R_2 = 2.7$ kΩ, $R_3 = 4.7$ kΩ and $V_{CC} = +11$ V. Assuming that the reactance of capacitor (C) is zero at frequencies of interest and that the transistor has a β of 100, calculate the voltage gain and input resistance of the circuit.

Linear integrated circuit techniques

5

☐ To describe the operation of a differential amplifier in terms of both its d.c. transfer characteristic and its small-signal a.c. performance.

☐ To explain the significance of the terms *differential gain*, *common-mode gain* and *common-mode rejection ratio*.

☐ To design current sources and use current mirror circuits.

☐ To explain the significance of being able to use active devices as loads instead of resistors.

☐ To design level shifting circuits including the amplified diode (V_{BE} multiplier).

☐ To explain the circuit function of a differential comparator and, by application of positive feedback, obtain hysteresis.

Objectives

This chapter aims to introduce techniques that are relevant to the design of linear integrated circuits yet widely applicable to circuit design using discrete components and arrays of multiple transistors.

The intent is to make use, wherever possible, of the precision matching of integrated transistors and resistor ratios and their tracking with temperature, and furthermore to remove capacitors, both coupling and decoupling, from circuit configurations and to rely on direct-coupling techniques. Under these circumstances, the differential amplifier is very important, providing, as well as a high-voltage gain, the subtracting (or differencing) function so vital to the ubiquitous operational amplifier.

Treatment of circuitry using operational amplifiers is outside the scope of this text. Much useful detail can be found in Millman and Grabel (1987), Chapters 10 and 16, Clayton (1979), and Horrocks (1990), Chapters 6, 7, 8 and 9.

The differential amplifier

The differential amplifier (Fig. 5.1) comprises two identical transistors coupled in a symmetrical manner at their emitters with d.c. bias current provided by a constant current source (I_T). Input voltages (V_1 and V_2) are applied at the bases of the two transistors while output currents are available at their collectors; conversion to output voltages is achieved by incorporating equal collector load resistors (R_C) connected to the positive supply voltage (V_{CC}).

With reference to Fig. 5.1,

$$V_1 - V_{BE1} = V_2 - V_{BE2}$$

Therefore

$$V_1 - V_2 = V_{BE1} - V_{BE2} \qquad (5.1)$$

and

$$I_{E1} + I_{E2} = I_T \qquad (5.2)$$

A general reference on the differential amplifier is Millman and Grabel (1987), pp. 435–441.

The close thermal tracking of integrated circuit components improves the performance of the differential amplifier.

Note that the transistor pair is biased by a current source.

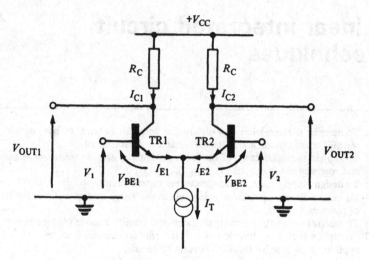

Fig. 5.1 Differential amplifier.

Balance state

Since the circuit is fundamentally symmetrical, at the balance condition of equal input voltages the bias current splits equally between the two transistors, i.e. if $V_1 = V_2$, Equation 5.1 gives $V_{BE1} = V_{BE2}$ and hence

$$I_{E1} = I_{E2} = \frac{I_T}{2} \tag{5.3}$$

If the current gains are identical,

$$I_{C1} = I_{C2} \tag{5.4}$$

and

$$V_{OUT1} = V_{OUT2}$$

Hence, for a zero differential input, the differential output is zero.

Large-signal behaviour

The balance is disturbed by, say, keeping V_2 constant and varying V_1. Increasing V_1 above V_2 causes V_{BE1} to exceed V_{BE2} (Equation 5.1). Therefore, TR1 conducts more heavily than TR2 but the sum of the two emitter currents is constrained to be constant by the bias current source (Equation 5.2). Thus, I_{C1} increases at the expense of I_{C2}.

Similarly, if V_1 is decreased below V_2, conduction in TR1 is reduced while the current in TR2 increases. Under the action of input voltage changes around the balance state, the d.c. bias current is steered towards one transistor or the other depending on the sense of the difference between the input voltages.

The corresponding collector currents (and voltages generated) are measures of the

The d.c. bias current is often referred to as the tail current, and the differential amplifier as a long-tailed pair.

difference between the input voltages V_1 and V_2. No implication of a linear relationship between input and output can be made; certainly, the nonlinear device characteristics would render such a relationship unlikely.

To analyse the d.c. (or large-signal) behaviour of the differential amplifier we invoke a description of I_E as a function of V_{BE} (Equation 1.17), which takes the form of the familiar diode equation:

Large-signal analysis takes into account the nonlinearity of a circuit.

$$I_E \approx I_s \exp\left(\frac{qV_{BE}}{kT}\right)$$

where, as before, I_s is a leakage or saturation current, q is electronic charge, k is Boltzmann's constant and T is temperature. For TR1

$$I_{E1} = I_{s1} \exp\left(\frac{qV_{BE1}}{kT}\right)$$

and for TR2

$$I_{E2} = I_{s2} \exp\left(\frac{qV_{BE2}}{kT}\right)$$

If the two transistors are assumed to be identical and at the same temperature,

$$I_{s1} = I_{s2} = I_s$$
$$T_1 = T_2 = T$$

and

$$\alpha_1 = \alpha_2 = \alpha$$

Now

$$I_T = I_{E1} + I_{E2}$$

$$= I_s\left[\exp\left(\frac{qV_{BE1}}{kT}\right) + \exp\left(\frac{qV_{BE2}}{kT}\right)\right]$$

$$\frac{I_{E1}}{I_T} = \frac{I_s \exp\left(\frac{qV_{BE1}}{kT}\right)}{I_s\left[\exp\left(\frac{qV_{BE1}}{kT}\right) + \exp\left(\frac{qV_{BE2}}{kT}\right)\right]}$$

$$= \frac{1}{1 + \exp\left(\frac{q(V_{BE2} - V_{BE1})}{kT}\right)} \tag{5.5}$$

Since $I_{C1} = \alpha I_{E1}$ and $V_{BE2} - V_{BE1} = V_2 - V_1$,

$$I_{C1} = \frac{\alpha I_T}{1 + \exp\left(\frac{q(V_2 - V_1)}{kT}\right)} \tag{5.6}$$

Similarly,

$$I_{C2} = \frac{\alpha I_T}{1 + \exp\left(\frac{q(V_1 - V_2)}{kT}\right)} \tag{5.7}$$

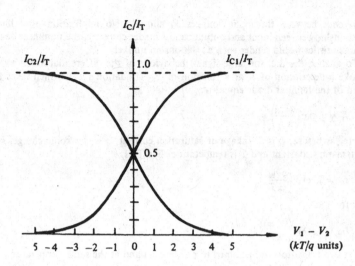

Fig. 5.2   Transfer characteristic of differential amplifier.

One quick check on the analysis!

If $V_1 = V_2$, then $I_{C1} = I_{C2} = 0.5 \times \alpha \times I_T$ which is the balance state.

Equations 5.6 and 5.7 are a close approximation to the large-signal d.c. behaviour of the differential amplifier and are plotted in Fig. 5.2. The nonlinear relationship between collector current and differential input voltage is obvious but several useful observations can be made:

1. The transfer characteristic is approximately linear over the differential input voltage range $\pm kT/q$ (i.e. approximately 50 mV peak-to-peak).

Clayton (1975), Chapters 4, 5 and 6, presents an illuminating discussion about applications.

2. The slope of the transfer curves is dependent on the tail current. The gain (transconductance, to be accurate) can be changed by altering I_T without affecting linearity in the central region. This effect is used in linear multipliers which multiply together two analogue signals, a very useful function for communication and instrumentation systems.

Millman and Grabel (1987), pp. 251–260 provide further details on ECL.

3. The differential amplifier is a natural limiter; there are no significant output changes for input excursions exceeding $\pm 4kT/q$ or ± 100 mV. This is the basis of current-steering logic circuits such as emitter-coupled logic (ECL).

Single-ended and **double-ended** outputs.

4. Output may be taken from the collector of either transistor (this is called **single-ended output**) or differentially between the two collectors (as $V_{OUT1} - V_{OUT2}$), which is called **double-ended output**.

Inverting and **non-inverting** inputs. The same terminology is applied to the inputs of an operational amplifier which has a single output. + and − signs are often applied to the non-inverting and inverting inputs respectively.

5. If V_1 increases relative to V_2, I_{C1} increases and V_{OUT1} falls. Therefore, with respect to output V_{OUT1}, the base of TR1 is an inverting input and the base of TR2 is a non-inverting input. The rôles of the inputs are reversed when referenced to output V_{OUT2}.

In practice, even a small mismatch between transistors or resistors causes the output to respond to common-mode input.

6. The output signal is a function only of the differential input signal and the tail current (I_T). If the bases of the two transistors are connected together, the circuit is in its balanced state, i.e. the transistors share the tail current equally. This balance is not disturbed by varying the base voltage, termed the **common-mode input signal**, and if the tail current is constant, the collector currents and voltages do not change.

No capacitors are required in the differential amplifier circuit – connection is made directly to the inputs. One input may be connected to a reference potential which is set at the d.c. level of the other (signal) input. Usually d.c. feedback techniques are used to stabilize the operating conditions of the overall circuit.

Discrete transistors are unsuitable for amplification of d.c. signals even using a differential amplifier because imbalance due to mismatch and temperature drift can mask the effect of an applied signal. Perfect balance is achieved for zero differential input voltage if the transistors are matched and if their V_{BE}s and current gains track with temperature; all properties of integrated transistors. However, since the voltage axis of the transfer characteristic is scaled in kT/q units, when the differential amplifier is unbalanced, its transconductance is temperature dependent.

Dual transistors, or arrays comprising several transistors all closely matched, are available for use in differential amplifiers and similar circuits which require matched devices in close thermal proximity.

Exercise 5.1

By differentiating Equation 5.6, prove that the differential transconductance (g_{md}) is given by

$$g_{md} = \frac{dI_{C1}}{d(V_1 - V_2)} = \frac{q\alpha I_T}{4kT}$$

which is half of the g_m of a single transistor operating at a d.c. emitter current of $I_T/2$.

Further consideration of the differential amplifier is directed towards its small-signal performance in terms of voltage gain, quality of differencing action and the application of negative feedback to the circuit.

Small-signal behaviour

Analysis of the differential amplifier for a.c. small signals in the region of the balance state provides the equivalent circuit of Fig. 5.3 using the simple BJT model developed in Chapter 3. Identical devices at the same temperature with equal collector load resistors and an ideal tail current source (infinite output resistance) are assumed.

The d.c. tail current source is open-circuit and V_{CC} is equivalent to earth for a.c. analysis.

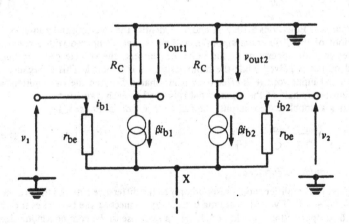

Fig. 5.3 Small-signal equivalent circuit of differential amplifier.

Summing currents at node X,

$$i_{b1} + \beta i_{b1} + i_{b2} + \beta i_{b2} = 0$$

Therefore

$$i_{b1} = -i_{b2} \tag{5.8}$$

Now

$$v_1 - v_2 = i_{b1}r_{be} - i_{b2}r_{be}$$

$$= (i_{b1} - i_{b2})r_{be}$$

$$= 2i_{b1}r_{be} \tag{5.9}$$

Also

$$v_{out1} = -R_C \beta i_{b1}$$

$$= -\frac{\beta R_C}{2r_{be}}(v_1 - v_2)$$

Therefore

$$A_{V1} = \frac{v_{out1}}{v_1 - v_2} \approx -\frac{R_C}{2r_e} \tag{5.10}$$

Similarly

$$A_{V2} = \frac{v_{out2}}{v_1 - v_2} \approx +\frac{R_C}{2r_e} \tag{5.11}$$

From Equations 5.10 and 5.11, the double-ended output voltage gain is

$$\frac{v_{out1} - v_{out2}}{v_1 - v_2} \approx -\frac{R_C}{r_e}$$

The differential input resistance (r_{in}) is quickly derived from Equation 5.9 as

$$r_{in} = \frac{v_1 - v_2}{i_{b1}} = 2r_{be} \tag{5.12}$$

The results of Equations 5.10, 5.11 and 5.12 identify the inverting and non-inverting functions of the differential amplifier and show that, compared with a common-emitter amplifier operating at the same d.c. current, the voltage gain for single-ended output is halved while the input resistance is doubled. This is because the differential input voltage is effectively being shared between the two emitter-base junctions; only half of the input signal is applied to each transistor.

Taking a double-ended output, the full voltage gain is restored, i.e.

$$A_V = \frac{v_{out1} - v_{out2}}{v_1 - v_2} = -\frac{R_C}{r_e} \tag{5.13}$$

Common-mode performance

A differential amplifier should respond only to the difference between the two input signals ($v_1 \sim v_2$). If v_1 and v_2 are made equal by connecting the two bases together then the output voltage should not change in response to the common-mode signal ($v = v_1 = v_2$). Optimum differencing performance is produced in a differential

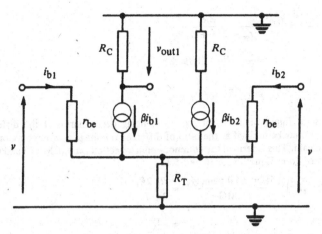

Fig. 5.4 Equivalent circuit of differential amplifier with non-ideal tail current source.

amplifier when the tail current source is ideal, i.e. I_T remains constant independent of the input signal level. Consider the case where, as an inexpensive expedient, I_T is derived using a resistor R_T connected to a negative supply voltage. I_T then varies with the applied input voltage. If the common-mode input voltage (v) is increased, the resultant increase in I_T is shared between the two transistors and both collector voltages fall. Therefore, in single-ended mode, there is an output response to the common-mode input and a common-mode voltage gain (CMG) can be defined as

If, in a circuit, a signal and its reference earth are both contaminated to an equal extent by interference (noise), this may be removed by connecting both signal and earth to the inputs of a differential amplifier which, ideally, responds to the signal and not to the common-mode interference.

$$\text{CMG} = \frac{v_{\text{out1}}}{v} \quad \text{or} \quad \frac{v_{\text{out2}}}{v} \tag{5.14}$$

In double-ended output mode, if the circuit is truly symmetrical, both collector voltages alter by the same amount and the differential output voltage is zero giving, in this case, a common-mode gain of zero.

Returning to the single-ended output mode, the circuit may be analysed to yield an expression for its common-mode gain. Consider the small-signal equivalent circuit of Fig. 5.4 with the finite tail current source resistance represented by a resistor R_T. Since the circuit is assumed symmetrical with common-mode input (v) applied,

$$i_{b1} = i_{b2}$$

Also,

$$v_{\text{out1}} = -\beta i_{b1} R_C$$

and

$$v = i_{b1} r_{be} + (1 + \beta) i_{b1} R_T + (1 + \beta) i_{b2} R_T$$

$$= i_{b1} r_{be} + 2(1 + \beta) i_{b1} R_T$$

Therefore, the voltage gain in respect of a common-mode input signal and single-ended output is given by

Common-mode voltage gain is also relevant to double-ended output but, in this case, should be zero for a perfectly symmetrical circuit.

$$CMG = \frac{v_{out1}}{v} = \frac{-\beta R_C}{r_{be} + 2(1 + \beta)R_T}$$

$$\approx \frac{-R_C}{r_e + 2R_T} \quad \text{if } \beta \gg 1$$

$$\approx \frac{-R_C}{2R_T} \quad \text{for } R_T \gg r_e \tag{5.15}$$

A useful figure of merit for the differencing performance of the differential amplifier may be defined as the ratio of differential voltage gain to common-mode voltage gain. This is termed the **common-mode rejection ratio** (CMRR) and may be calculated from Equations 5.10 and 5.15.

The CMRR is often expressed in decibels (dB) given by:

CMRR (dB) = 20 log$_{10}$ (CMRR)
20 dB ≈ 10 (ratio)
40 dB ≈ 100 (ratio)
100 dB ≈ 10^5 (ratio), etc.

$$CMRR = \frac{\text{Differential gain, } A_{V1}}{CMG} \approx \frac{R_C}{2r_e}\frac{2R_T}{R_C}$$

$$= \frac{R_T}{r_e} \tag{5.16}$$

This result proves conclusively that, for maximum rejection of common-mode signals, the resistance of the tail current source should be infinite.

Exercise 5.2

Remember that, in the quiescent balance condition, the tail current splits equally between the two transistors:

$I_{E1} = I_{E2} = 0.5I_T$

For a differential amplifier with a tail current of 5 mA and equal collector load resistors of 4.7 kΩ, calculate the single-ended voltage gain and the differential input resistance. (Assume $\beta = 100$ and that the tail current source is ideal.)
[*Answer*: $|A_v| = 235$, $r_{in} = 2.02$ kΩ]
If the current source is now derived using a 2.2 kΩ resistor, calculate the common-mode gain and common-mode rejection ratio.
[*Answer*: CMG = 1.066, CMRR = 220.5]

Emitter feedback

Series feedback may be incorporated in a differential amplifier by introducing equal resistors into each emitter (Fig. 5.5). As would be expected, input resistance is increased and voltage gain reduced, the latter being defined by a resistor ratio rather than the bias-dependent parameters of the transistors.

It is left as an exercise for the reader to prove that

$$A_{V1} = \frac{v_{out1}}{v_1 - v_2} \approx \frac{-R_C}{2(R + r_e)} \text{ if } \beta \gg 1 \tag{5.17}$$

and

$$r_{in} = 2[r_{be} + (1 + \beta)R] \tag{5.18}$$

where R is the resistance in each emitter lead.

Equation 5.17 shows that this series negative feedback has the effect of linearizing the transfer characteristic since its dependence on the nonlinear r_e is reduced. The input resistance is increased over that for the basic configuration without feedback. An alternative method for increasing input resistance yet preserving a high transconductance is to use Darlington-connected transistors in place of single devices. The highest input resistance is achieved by using field-effect transistors (see Chapter 7).

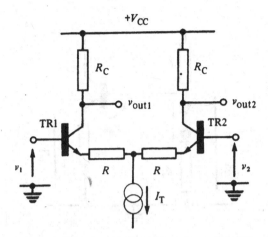

Fig. 5.5 Differential amplifier with emitter feedback.

Current sources

Current sources are very important electronic circuits providing, for example, biasing functions such as the tail current source of a differential amplifier or, when charging capacitors, means of generating non-sinusoidal waveforms such as ramps.

An ideal current source generates a fixed d.c. current independent of load impedance and applied voltage variations. It may be represented by a Norton generator with infinite output resistance. As we have already seen, the tail current source of a differential amplifier must have infinite output resistance to achieve an infinite CMRR or perfect differencing operation.

The simplest d.c. current source is achieved by using a single resistor whose value is calculated by applying Ohm's law to a knowledge of the current required and the d.c. voltage across the resistor. This passive current source has a Norton output resistance equal to the resistor value; consequently, to achieve high output resistance at moderate current levels, a very high supply voltage is required. For example, to achieve an output resistance of 1 MΩ with a current of 1 mA, a supply voltage of 1 kV must be provided – an expensive and dangerous exercise!

On the other hand, using an active current source such as a biased BJT, a very high output resistance is maintained down to a collector-emitter voltage of approximately 0.7 V, below which the onset of saturation dramatically reduces the output resistance. The voltage range over which a circuit behaves as a high-impedance current source is referred to as the **compliance** of the circuit.

Temperature stability

A BJT biased in the common-base or common-emitter configuration using one of the methods presented in Chapter 2 may be regarded as a current source controlling a d.c. current (I_C) flowing through the collector load resistor (R_C). Constancy or stability of I_C against transistor parameter variation with temperature is determined by the design of the bias circuit, the most important factor being the temperature dependence of the emitter-base voltage (approximately $-2\,\text{mV}/^\circ\text{C}$). This can be

If I is constant then

$$\frac{dv_c}{dt} = \frac{I}{C} = \text{constant}$$

Therefore $v_c\,(t)$ is a ramp.

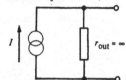

Remember that, in integrated circuits, resistor values are restricted to several tens of kilohms.

The upper limit of the compliance is determined by collector-base breakdown, beyond which the collector current is no longer, nominally, independent of collector voltage.

Leakage currents are also significant at high temperatures.

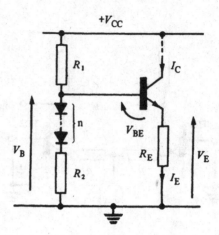

Fig. 5.6   Potentiometer bias circuit with temperature compensating diodes.

compensated by introducing diodes into the potentiometer bias chain as shown in Fig. 5.6.

For this circuit, $I_C \approx I_E = V_E/R_E$ and therefore is constant if V_E is constant. Assuming n diodes, each with a forward voltage drop V_D, and ignoring base current,

$$V_E = V_B - V_{BE}$$

$$= \frac{R_2}{R_1 + R_2}(V_{CC} - nV_D) + nV_D - V_{BE}$$

$$= \frac{R_2}{R_1 + R_2}V_{CC} + \left[\frac{R_1}{R_1 + R_2}nV_D - V_{BE}\right] \tag{5.19}$$

For temperature independence of V_E, and therefore of I_C, the bracketed term in Equation 5.19 must be zero, leaving

$$V_E = \frac{R_2}{R_1 + R_2}V_{CC} \tag{5.20}$$

If all the diodes are identical to the emitter-base junction and are in close thermal proximity to the transistor (conditions readily attainable in an integrated circuit or with integrated transistor arrays), then $V_D = V_{BE}$ at all temperatures and the number of diodes (n) determines the required R_2/R_1 ratio. For example, with two diodes, $n = 2$ and $R_2 = R_1$ from the bracket in Equation 5.19; and for three diodes, $n = 3$ and $R_2 = 2R_1$. In each case V_E is given by Equation 5.20 and R_E may be calculated to give the desired collector current.

An alternative method of achieving a temperature-stable BJT current source is to replace R_2 and the diodes of Fig. 5.6 with a breakdown diode which is selected to have a negative temperature equal to that of the emitter-base junction. With this arrangement, as temperature increases the base voltage decreases at the same rate as the base-emitter voltage, keeping constant the emitter voltage and hence the collector current. This tracking with temperature can be achieved using a diode with a nominal breakdown voltage of 3.9 V but manufacturers' data should be consulted for accurate information since the temperature coefficient is a function of diode reverse current as well as of breakdown voltage. This technique is restricted to discrete component circuits since diodes with a breakdown voltage of this magnitude are not generally available within integrated circuits.

Output resistance

The output resistance of an active current source would be expected to be infinite if the idealized output characteristic presented in Chapter 1 is accepted. However, this characteristic does not indicate the significant variation of I_C with V_{CE} which occurs in practice and is represented principally by r_{ce} in the hybrid-π model or by h_{oe} in the h-parameter description.

I_C can change by approximately 25% over the compliance range.

The output resistance of a BJT in common-emitter is given by

$$r_{out} \approx r_{ce} = \mu r_e \qquad (5.21)$$

where

$$10^3 < \mu < 10^5$$

and

$$r_e \approx \frac{25}{I_C} \quad (I_C \text{ in mA})$$

Taking $\mu = 5 \times 10^3$ and $I_C = 1\,\text{mA}$, the typical output resistance of a common-emitter circuit is calculated as 125 kΩ. Note that this figure is greater for lower currents (owing to the inverse proportionality between r_e and I_C) and is independent of the transistor β.

The expression of Equation 5.21 refers to a common-emitter circuit where the emitter is connected directly to earth (or to a voltage supply) or the emitter resistor (R_E) is bypassed for a.c. signals by a decoupling capacitor. It has been shown in Chapter 4 that, if the decoupling capacitor is removed, the resultant series negative feedback significantly increases the output resistance to

$$r_{out} = r_{ce}\left(1 + \frac{\beta R_E}{R_E + r_{be}}\right) + R_E \| r_{be} \qquad (5.22)$$

For example, if $I_C = 1\,\text{mA}$, $\beta = 100$ and $R_E = 2.2\,\text{k}\Omega$ (and taking $\mu = 5 \times 10^3$), r_{out} is now approximately 6 MΩ.

However, Equations 5.21 and 5.22 are both valid only if the external base resistance (R_B) is zero or shorted by an input signal voltage source. If the current source is potentiometer biased with a resistive bias chain, $R_B = R_1 \| R_2$ (see Fig. 2.7) modifies the output resistance to

The reader should prove this relationship; another chance for more practice at analysis!

$$r_{out} = r_{ce}\left(1 + \frac{\beta R_E}{R_E + r_{be} + R_B}\right) + R_E \| (r_{be} + R_B) \qquad (5.23)$$

a reduced figure compared to Equation 5.22.

Using Equation 5.23, calculate r_{out} if $I_C = 1\,\text{mA}$, $\beta = 100$, $R_E = 2.2\,\text{k}\Omega$ and $R_B = 5\,\text{k}\Omega$. ($\mu = 5 \times 10^3$)
[*Answer*: 2.96 MΩ]

Exercise 5.3

Three alternative biasing arrangements can be applied to recover the higher output resistance of Equation 5.22 with $R_B \approx 0$.

A discrete component technique; avoid capacitors in integrated circuit design.

1. Incorporate a decoupling capacitor from the transistor base to earth. However, this raises the output resistance only under a.c. conditions; at d.c. the output resistance is the lower figure given by Equation 5.23.
2. Connect the transistor base directly to a convenient d.c. supply voltage already provided in the system. This does not permit cancellation of the V_{BE} temperature coefficient unless a temperature-sensitive emitter resistor is used.
3. Use the breakdown diode biasing technique already discussed in this section. The low slope resistance (tens of ohms) of the diode, when incorporated as R_B in Equation 5.23, proves insignificant.

Sources or sinks?

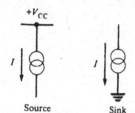

Source   Sink

A biased n–p–n bipolar transistor extracts a nominally constant collector current from the $+ V_{CC}$ supply and, to be correct, should be referred to as a **current sink** since it sinks current from a higher supply potential. The term **current source** should be restricted to apply only to circuits which control current flowing to a load (or other circuitry) at a lower potential. This description is appropriate to a biased p–n–p transistor. However, the generic term current source is commonly applied to both sources and sinks.

Current mirrors

Basic current mirror

Millman and Grabel (1987), pp. 397–402 describe current mirrors.

A very useful circuit which relies on the thermal tracking properties of matched BJTs in an integrated circuit is the current mirror (also termed the current repeater, or current mimic) which takes a source current and converts it to a precisely ratioed sink current.

In the basic current mirror circuit, shown in its n–p–n version in Fig. 5.7a, an input reference current $I_1 = (V_{CC} - V_{BE1})/R$ is applied to the diode-connected BJT (TR1) establishing a base-emitter voltage (V_{BE1}) appropriate to I_1. The base-emitter voltage (V_{BE2}) of transistor TR2 is forced to be equal to V_{BE1}. If the two transistors are matched and have infinite βs, the sink current flowing in TR2 is equal to the source current in TR1, i.e. $I_2 = I_1$.

A current mirror circuit comprising p–n–p transistors operates in exactly the same way except that the current directions are reversed and a current source results.

If the β of the transistors is finite then further analysis, annotated on Fig. 5.7a and considering the matched transistors as β-devices, i.e. $I_C = \beta I_B$, shows that the current mirror ratio is given by

$$\frac{I_2}{I_1} = \frac{\beta}{\beta + 2} \tag{5.24}$$

(In such an analysis, start with the lowest current level in the circuit (invariably a base current), set that current equal to one unit and repeatedly apply $I_C = \beta I_B$ for the transistors. Here $I_1 = (\beta + 2)$ current units while $I_2 = \beta$ current units and Equation 5.24 results.)

The attractive features of the basic current mirror lie in its simplicity and the fact that only one base-emitter voltage drop is required, thus normally releasing most of

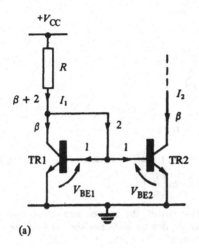

(a)

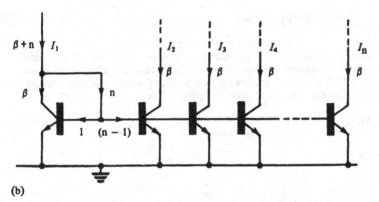

(b)

Fig. 5.7 (a) Basic current mirror. (b) Multiple repeater circuit.

the available supply voltage for other purposes. However, the circuit possesses only a moderately high output resistance (approximately r_{ce} of TR2) and the current mirror ratio is sensitive to transistor β (particularly for low values of β as shown by Equation 5.24) and also to offset between the devices.

The collector current (I_C) of a BJT can be related to its base-emitter voltage (V_{BE}) and a saturation current (I_s) by Equation 1.17

$$I_C \approx I_s \exp \frac{qV_{BE}}{kT}$$

The base-emitter offset (or mismatch) voltage between two transistors is defined as the difference in the base-emitter voltages which must be applied to achieve equal collector currents. Using subscripts 1 and 2 to distinguish between the two transistors which are assumed to be at the same temperature (T),

$$I_{C1} = I_{s1} \exp \left(\frac{qV_{BE1}}{kT} \right)$$

and

$$I_{C2} = I_{s2} \exp \left(\frac{qV_{BE2}}{kT} \right)$$

Therefore

$$\frac{I_{C1}}{I_{C2}} \frac{I_{s2}}{I_{s1}} = \exp \left[\frac{q}{kT} \left(V_{BE1} - V_{BE2} \right) \right] \tag{5.25}$$

For equal currents, i.e. $I_{C1} = I_{C2}$, the offset voltage ($\triangle V_{BE}$) is given by

$$\triangle V_{BE} = V_{BE1} - V_{BE2} = \frac{kT}{q} \ln \left(\frac{I_{s2}}{I_{s1}} \right) \tag{5.26}$$

If, as in the current mirror circuit, the base-emitter voltages are forced to be equal, any mismatch between the devices manifests itself as a difference in collector currents which may be calculated from Equations 5.25 and 5.26

$$\frac{I_{C1}}{I_{C2}} = \frac{I_{s1}}{I_{s2}} = \exp \left[\frac{q}{kT} \triangle V_{BE} \right] \tag{5.27}$$

where $\triangle V_{BE}$ is the base-emitter offset voltage.

Current dependence on V_{BE} mismatch is shown in Table 5.1 calculated, at $T = 290\,\mathrm{K}$, for several different values of offset voltage. It is significant that an offset of only 1 mV produces a current ratio 4% in error from the ideal case, further compounded by the $\beta/(\beta + 2)$ factor of Equation 5.24.

Table 5.1

$\triangle V_{BE}$ (mV)	$\dfrac{I_{C1}}{I_{C2}}$	error (%)
0.1	1.004	0.4
0.5	1.020	2.0
1.0	1.041	4.1
2.0	1.083	8.3
5.0	1.222	22.2

Integrated BJTs usually are guaranteed to have their V_{BE}s matched to within 5 mV.

Extra mirror transistors can be added as shown in Fig. 5.7b to provide (in n–p–n form) multiple current sinks. If n transistors are used including the diode-connected reference device it can be shown that, in the absence of mismatch error,

$$I_2, I_3, \ldots, I_n = I_1 \frac{\beta}{\beta + n} \tag{5.28}$$

The effect of mismatch can be reduced at the expense of assigning a larger fraction of the supply voltage to the current mirror by introducing equal resistors into each emitter ($R_1 = R_2$ in Fig. 5.8). Any offset between the BJTs can be made insignificant compared to the voltage dropped across the emitter resistors. Inclusion of these resistors also raises the output resistance due to series negative feedback.

Up to this stage we have considered the generation of a current equal to the

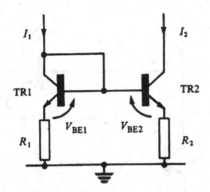

Fig. 5.8 Mirror circuit including emitter resistors.

reference current. It is possible to generate integer current ratios by parallelling transistors; for example, in Fig. 5.7b, if the collectors of the identical devices TR2 and TR3 are joined together, $I_2 + I_3 \approx 2I_1$.

Non-integer ratios can be achieved by recognizing that it is the current densities (current per unit emitter area) that are forced to be equal with equal V_{BE}s. For example, if transistor TR2 in Fig. 5.7a has three times the emitter area of TR1, then $I_2 \approx 3I_1$ since, assuming infinite βs,

$$\frac{I_2}{I_1} = \frac{\text{Emitter area of TR2}}{\text{Emitter area of TR1}}$$

Alternatively, a single resistor (R) may be included in either emitter circuit. The current ratio determines the difference between the base–emitter voltages appearing across R, from which R may be determined. For the case of Fig. 5.8, assuming infinite βs, matched transistors and $R_1 = 0$,

$$V_{BE1} = V_{BE2} + I_2R_2$$

Therefore

$$V_{BE1} - V_{BE2} = I_2R_2 = \frac{kT}{q}\ln\left(\frac{I_1}{I_2}\right) \qquad (5.29)$$

An interesting temperature transducer using current mirrors is described in Bannister and Whitehead (1991).

The temperature stability of such a circuit is complex since the resistor itself has a temperature coefficient and optimum results are usually achieved by purely active (all transistor) mirror circuits.

In the circuit of Fig. 5.8, $I_1 = 100\,\mu\text{A}$ and $I_2 = 10\,\mu\text{A}$. Assuming matched transistors, calculate the required value of R_2 at a temperature of 17°C. ($R_1 = 0$.) [Answer: $R_2 = 5.76\,\text{k}\Omega$]

Exercise 5.4

Wilson mirror

A more sophisticated current mirror circuit, the Wilson mirror shown in Fig. 5.9, reduces the current ratio dependence on β and raises the output resistance with the penalty of only one extra transistor and one extra V_{BE} voltage drop across the

A different form of three-transistor current mirror circuit is described in Millman and Grabel (1987), pp. 400–401.

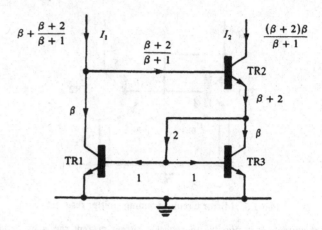

Fig. 5.9   Wilson current mirror circuit.

circuit. Analysis (again starting with the lowest current and simply considering the transistors as β devices) yields, for the Wilson mirror

$$\frac{I_2}{I_1} = \frac{\beta^2 + 2\beta}{\beta^2 + 2\beta + 2} \tag{5.30}$$

Output resistance is increased by series feedback, the diode-connected transistor TR3 in the emitter of TR2 giving an output current I_2 less dependent on the collector voltage of TR2 than in the basic current mirror circuit.

Exercise 5.5   To show the reduction in β dependence of the Wilson mirror circuit, calculate the I_2/I_1 ratios of Equations 5.24 and 5.30 for βs of 10 and 100.
[*Answer*: Basic current mirror: $I_2/I_1 = 0.833$ ($\beta = 10$), 0.98 ($\beta = 100$). Wilson current mirror: $I_2/I_1 = 0.984$ ($\beta = 10$), 0.9998 ($\beta = 100$).]

Active loads

The economic unit in integrated circuit design is silicon area as increasing the area reduces yield and consequently increases cost.

In designing integrated circuits the economic rules of discrete component circuit design are overturned. Active devices (transistors and diodes) are inexpensive since they are of much smaller area than resistors or capacitors. Hence, every effort must be made to minimize the total resistance in a circuit and certainly replace passive components with transistors wherever possible.

The differential amplifier is a valuable circuit since no capacitors are necessary and the tail current source can be generated by a current mirror comprising two transistors and a resistor. Let us now replace the collector load resistors with several diode-connected transistors (Fig. 5.10). For n diodes the equivalent slope resistance is nr_e. Hence the single-ended small-signal voltage gain is given by

$$|A_v| = \frac{nr_e}{2r_e} = \frac{n}{2} \tag{5.31}$$

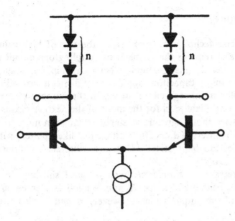

Fig. 5.10 Differential amplifier with diode loads.

Only a limited number of diodes can be incorporated as a load since each diode drops approximately 0.7 V when conducting. Therefore, the voltage gain is low in value but the cancellation of the nonlinear r_es results in a nominally linear voltage gain with little distortion.

The full differential capability of the differential amplifier can be realized by using a p–n–p current mirror as an active load as shown in Fig. 5.11. This load circuit is very economical in area since TR1 provides the current I_1 for the current mirror and no resistors are used. Since $I_3 = I_1$ (ideally), the output current I_o is $I_1 - I_2$ at an impedance level approximately equal to r_{ce} of transistor TR4 in parallel with the output resistance of TR2. Neglecting the loading effect of subsequent circuitry, high voltage gains can be achieved between the differential input and the output.

It is interesting that one nonlinearity can be cancelled by another.

See Millman and Grabel (1987), pp. 611–616.

It can be shown that the output resistance of TR2 is $2r_{ce}$. (Another opportunity to practise analysis!)

$$A_V = \frac{2r_{ce} \; /\!/ \; r_{ce}}{2r_e} = \frac{\mu}{3} \approx 1500$$

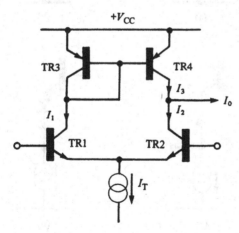

Fig. 5.11 Differential amplifier with current mirror load.

Level-shifting circuits

Despite using circuit techniques which avoid the use of high-valued decoupling capacitors, there still remains the problem of coupling one circuit to another. As illustrated in Chapter 2, direct coupling between amplifier stages removes the requirement for coupling capacitors but it is very difficult, sometimes impossible, to design a complex multistage circuit in which the output voltage of each stage corresponds to a convenient level for the input of the next. A means of shifting the d.c. level of signals without introducing signal attenuation must be found.

The ideal coupling inserts a d.c. offset voltage just like a battery with zero internal resistance (batteries are impractical within even discrete component circuits, let alone integrated circuits!).

Fig. 5.12 illustrates four simple methods of level shifting. In the circuit of Fig. 5.12a, the split emitter resistance (R_1 and R_2) acts as a potential divider lowering the d.c. level of the output signal. However, R_1 and R_2 also attenuate the a.c. signal.

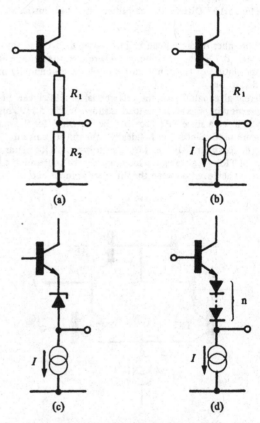

Fig. 5.12   Various level-shifting circuits.

If R_2 is replaced by a constant current source (Fig. 5.12b), which has a high output resistance, the attenuation now is minimal and the d.c. voltage shift is readily calculated as IR_1.

The current source can be a transistor.

R_1 can be replaced by a breakdown diode but this technique is not favoured in integrated circuit design since the only practical device obtainable without special processing is a reverse biased emitter-base junction which has a breakdown voltage of approximately 7 V. Unfortunately breakdown diodes are noisy devices; the breakdown mechanism is a statistical process rather than a steady one and the voltage across a breakdown diode, in breakdown, is subject to random fluctuations (noise) with an amplitude in the order of millivolts. In a linear circuit much care is taken in design to minimize the amount of noise added to, and hence degrading, the desired signal. For this reason, breakdown diode coupling in linear circuits is seldom used although digital circuits can tolerate this technique if a nominal 7 V offset is required.

Alternatively a series chain of n forward biased diode-connected transistors (Fig. 5.12d) can be used to introduce a d.c. level shift of approximately $n \times 0.7$ V. The series impedance of this coupling is nr_e which can be made small by operating at a high forward current (mA). This technique, however, is restricted to a level shift of an integral number of forward diode voltage drops.

The amplified diode

A very effective method of generating a voltage offset greater than 0.7 V with a minimal number of components is by using the amplified diode circuit of Fig. 5.13a. A current I flows through the network consisting of TR1, R_1 and R_2. TR1 conducts and its V_{BE} is set by its emitter current I_E. If the β of the transistor is high, I_B may be neglected and the current $I' = (I - I_E)$ flows through R_1 and R_2. Since the voltage across R_1 is

This circuit is widely used in audio power amplifiers (see Chapter 8).

$$V_{R1} = I'R_1 = V_{BE}$$

the voltage across R_2,

$$V_{R2} = I'R_2 = \frac{R_2}{R_1} V_{BE}$$

Therefore, the total voltage (V) across the circuit is given by

$$V = V_{R1} + V_{R2} = \left(1 + \frac{R_2}{R_1}\right) V_{BE} \tag{5.32}$$

V is then a multiple (which may be non-integer depending on the ratio of R_1 and R_2) of the base-emitter voltage of TR1, hence the term **amplified diode**. V is also temperature dependent; differentiating Equation 5.32 with respect to temperature gives

Another name for the amplified diode is the V_{BE} multiplier.

$$\frac{dV}{dT} = \left(1 + \frac{R_2}{R_1}\right) \frac{dV_{BE}}{dT} \tag{5.33}$$

again resistor ratio determined, a feature which is highly desirable in some circuits, for example in biasing the output stage of a power amplifier. In discrete component circuits the resistor R_2 is usually made adjustable to overcome component tolerances.

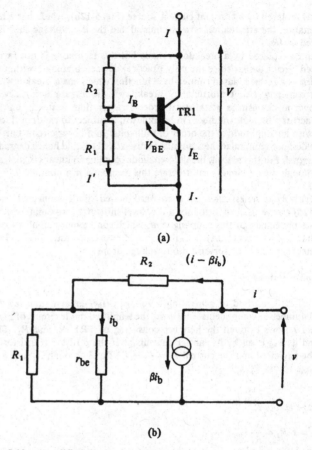

(a)

(b)

Fig. 5.13 Amplified diode circuit (a) with its a.c. equivalent circuit (b).

Let us now analyse the equivalent circuit of Fig. 5.13b to generate an expression for the slope resistance of the amplified diode circuit.

$$i_b = \frac{R_1}{R_1 + r_{be}} (i - \beta i_b)$$

Therefore

$$i_b \left(1 + \frac{\beta R_1}{R_1 + r_{be}} \right) = \frac{R_1}{R_1 + r_{be}} i$$

and

$$i_b = i \left(\frac{R_1}{r_{be} + (1 + \beta) R_1} \right)$$

Now

$$v = (i - \beta i_b) [R_2 + (R_1 \| r_{be})]$$

Substituting for i_b gives

$$v = i \left[1 - \frac{\beta R_1}{r_{be} + (1 + \beta) R_1} \right] \left[R_2 + \frac{R_1 r_{be}}{R_1 + r_{be}} \right]$$

$$= i \left[\frac{r_{be} + R_1}{r_{be} + (1 + \beta) R_1} \right] \left[\frac{R_1 R_2 + r_{be} (R_1 + R_2)}{R_1 + r_{be}} \right]$$

slope resistance,

$$r = \frac{v}{i} = \frac{R_1 R_2 + r_{be} (R_1 + R_2)}{r_{be} + (1 + \beta) R_1} \qquad (5.34)$$

$$\approx \frac{R_2}{\beta} + \left(1 + \frac{R_2}{R_1} \right) r_e \qquad (5.35)$$

if $\beta \gg 1$ and $R_1 \gg r_e$.

Unfortunately this is not a simple expression which can be used in design. However, it permits calculation of the slope resistance of the amplified diode circuit once current levels and the resistor values have been determined.

Exercise 5.6

Calculate the d.c. terminal voltage and the a.c. slope resistance of the amplified diode circuit of Fig. 5.13 in which $I = 2.7$ mA, $R_1 = 1$ kΩ and $R_2 = 4.7$ kΩ. The transistor β is 100 and V_{BE} is 0.7 V. (Hint: $I_E = 2$ mA.)
[Answer: $V = 3.99$ V and $r = 116$ Ω]

The slope resistance of the amplified diode is always greater than that of the equivalent number of diode-connected transistors required to give a specified voltage offset; the second term in Equation 5.35 corresponds to the slope resistance of the diodes, leaving the first term as the difference between the two configurations. However, the value of the amplified diode circuit lies in its ability to provide a nonintegral number of forward diode voltage drops with a temperature dependence which is a multiple of that of a single diode.

A simple operational amplifier

Figure 5.14 shows a very simplified schematic of several integrated circuit techniques combined to form an operational amplifier circuit. Transistors TR1 and TR2 form the differential input stage with an active load (TR3, TR4). The output from this stage is amplified by the Darlington pair (TR5, TR6) with constant current load (TR7) being a section of the multiple current mirror (TR7, TR8, TR9) which also provides the tail current for TR1 and TR2. The output stage comprises complementary (n-p-n and p-n-p) emitter followers (TR11 and TR12) biased into conduction by the amplified diode (TR10).

Millman and Grabel (1987, Chapter 14) describe in some detail the configuration of the popular type 741 operational amplifier.

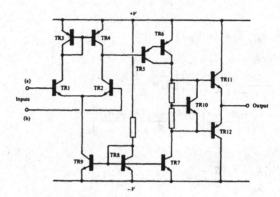

Fig. 5.14   A very simplified schematic of an operational amplifier illustrating techniques encountered in this chapter.

The differential comparator

Millman and Grabel (1987, Chapter 15) and Clayton (1979, Chapter 7) discuss comparators and some applications.

The differential comparator, usually classified as a linear integrated circuit, is widely used in interfacing between analogue and digital circuitry. Its function is to compare an input signal voltage with a reference voltage (which may be a second input signal) and provide a digital output (two-state, high or low logic levels) depending on the

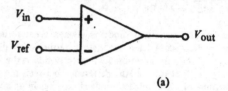

(a)

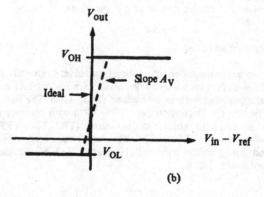

(b)

Fig. 5.15   Differential comparator (a) and its transfer characteristic (b).

sense of the inequality between the two inputs. The output is directly compatible with logic families such as TTL, ECL and CMOS, while the inverting ($-$) and non-inverting ($+$) inputs have the same significance as for differential and operational amplifiers. (An operational amplifier can also be used as a comparator provided its output voltage limits are suitably constrained.) Referring to Fig. 5.15, if V_{in} is greater than V_{ref} then V_{out} is at the high logic level (V_{OH}) and, when V_{in} is less than V_{ref}, V_{out} is at the low logic level (V_{OL}).

An ideal comparator has infinite voltage gain (A_V) producing an abrupt change of output state when the input voltage inequality is reversed. In practice, voltage gains range from 1500 to 500 000 and output transitions may be as long as 1.3 μs or as short as 2 ns. The static voltage characteristic is shown in Fig. 5.15b.

The basic comparator suffers from two disadvantages. First, if the rate of change of the input voltage through the reference level is very slow, the rate of change of output voltage (given by A_V times the input rate of change) is slow and may not meet the input speed specification of the digital circuit driven by the comparator. Second, the input signal is always contaminated by noise to a lesser or greater extent. As the nominal input signal passes through the reference level, superimposed noise

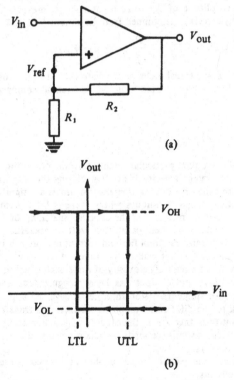

(a)

(b)

Fig. 5.16 (a) Hysteresis applied to a comparator. (b) The modified transfer characteristic.

causes multiple output transitions which can lead to system malfunction, especially when the comparator is driving a digital counter. Both of these disadvantages can be removed by applying positive feedback to the basic comparator circuit (Fig. 5.16a). The input is connected to the inverting input terminal. With V_{in} more negative than any circuit potential, $V_{out} = V_{OH}$ and, assuming that the non-inverting input does not load the circuit,

This circuit is also known as a Schmitt trigger.

$$V_{ref} = V_{OH} \frac{R_1}{R_1 + R_2} = \text{Upper threshold level (UTL)} \tag{5.36}$$

If V_{in} is now increased to reach this upper threshold level, V_{out} commences to fall, progressively reducing V_{ref} below V_{in} and driving V_{out} very rapidly towards V_{OL}. When $V_{out} = V_{OL}$,

$$V_{ref} = V_{OL} \frac{R_1}{R_1 + R_2} = \text{Lower threshold level (LTL)} \tag{5.37}$$

A further change of output level cannot occur until V_{in} falls to this lower threshold level. The voltage difference between the threshold levels (UTL − LTL) is termed the **hysteresis** (or dead-band, or backlash) of the circuit which, if designed to be greater than the amplitude of the noise input voltage, prevents multiple output transitions. The hysteresis is determined from Equations 5.36 and 5.37

$$\text{Hysteresis} = \text{UTL} - \text{LTL} = \frac{R_1}{R_1 + R_2} (V_{OH} - V_{OL}) \tag{5.38}$$

and modifies the voltage transfer characteristic to that of Fig. 5.16b.

In Chapter 6 there are two applications of differential comparators to timing circuits.

Summary

In this chapter we have paid particular attention to the constraints imposed upon circuit design by the integrated circuit process. Noting that capacitors and high-valued resistors are expensive in area whereas transistors are significantly smaller, we have investigated design techniques where there is little premium placed on transistor count. Further, recognizing the superior matching of integrated components and the tracking of their parameters with temperature variations, the unique value of the differential amplifier and current mirrors can be realized.

A differential amplifier with its subtracting function is vital as the input stage of an operational amplifier circuit. The quality of differencing action compared with its response to common-mode input has been recognized as a measure of its performance. As with single-transistor amplifier stages, the application of series negative feedback to the differential amplifier linearizes its transfer characteristic and enhances input resistance but at the expense of a reduction in gain. Also, we have seen how to replace resistive loads by active devices, diode-connected transistors or current mirrors.

We have recognized the need for direct coupling and the consequent requirement for level-shifting circuits.

The operational amplifier is a supreme example of the economics of integrated circuits – a mass-produced, inexpensive building block with a myriad of applications. Although it is beyond the scope of this text to design a practical operational amplifier and even to investigate its applications, an understanding of the circuit elements fundamental to linear integrated circuit design benefits the designer of discrete component circuits.

Problems

5.1 Referring to the differential amplifier of Fig. 5.17, for the values of V_1 given in Table 5.2, calculate the values of V_2 and V_3. V is a fixed reference voltage of +2 V. Assume that the βs of the matched transistors are 100 and that the V_{BE} of a conducting transistor is 0.7 V. Also, identify each transistor state (e.g. cut-off, active, saturated) under all of the above conditions.

Table 5.2

V_1 (V)	V_2 (V)	V_3 (V)	TR1	TR2
1.5				
2.0				
2.5				
3.0				

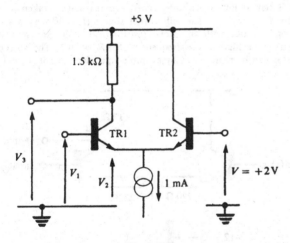

Fig. 5.17

5.2 Repeat Problem 5.1 for the case where the tail current source is replaced by a 1 kΩ resistor connected between the junction of the transistor emitters and earth.

5.3 Calculate the common-mode rejection ratio for the differential amplifier of Fig. 5.18 with both inputs in the vicinity of earth potential. Which is the non-

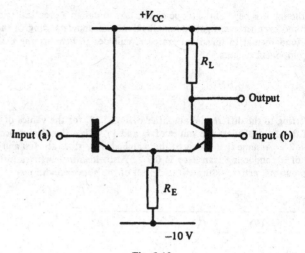

Fig. 5.18

inverting input? (Hint: Do not be discouraged by the absence of component values!)

5.4 In a differential amplifier a single-ended output voltage is taken across a 1 kΩ collector load resistor. The tail current source ($I_T = 10\,\text{mA}$) is a common-base configuration with an emitter resistor (R_E). Calculate the value of R_E necessary to achieve a low-frequency CMRR of 10^5. The current source transistor can be represented by a simple equivalent circuit involving r_{be}, r_{ce}

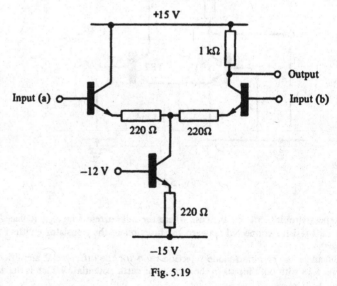

Fig. 5.19

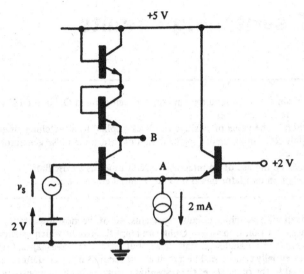

Fig. 5.20

and a current source βi_b. The β of all the transistors is greater than 30 and r_{ce} of the current source transistor is 50 kΩ.

5.5 For the differential amplifier circuit of Fig. 5.19:
(a) Which input is the inverting input?
(b) Calculate the CMRR at $V_a = V_b = 0$ V given that, for all transistors, $\beta = 50$ and $\mu = 5 \times 10^3$.
For the current source transistor assume a small-signal model incorporating r_{be}, r_{ce} and a current source βi_b.

5.6 Fig. 5.20 shows a differential amplifier with two diode-connected transistors in series forming a collector load. All transistors are identical with βs of 100 and it may be assumed that the V_{BE} of a conducting transistor is 0.7 V.
(a) What is the incremental resistance provided by the two load transistors?
(b) Sketch the voltage waveforms at points A and B in response to a small-signal input voltage (v_s).

5.7 In the current mirror circuit of Fig. 5.7 the transistors have a V_{BE} offset of 3 mV. If the βs of both devices are 20, calculate the current mirror ratio (I_2/I_1) at a temperature of 17 °C.

5.8 In the current mirror circuit of Fig. 5.8, $I_1 = 100\ \mu A$ and $I_2 = 5\ \mu A$. Calculate the required value of R_2 if $R_1 = 0$ and the temperature is 290 K.

5.9 An amplified diode circuit is required to develop a nominal voltage of 2.25 V when passing a minimum current of 1 mA. Assuming that the V_{BE} of a conducting transistor is 0.7 V and that base current can be ignored, calculate the values of the two resistors.

5.10 Positive feedback is applied to a differential comparator by resistors R_1 and R_2 as shown in Fig. 5.16. If $R_1 = 1$ kΩ and $R_2 = 4.7$ kΩ and the output voltage logic levels are $+ 3.5$ V and $- 0.5$ V, determine the upper and lower threshold levels and the hysteresis of the circuit.

6 BJT switching circuits

Objectives
- [ ] To explain the drive requirements for the well defined ON and OFF states of saturating BJT switches.
- [ ] To describe the value of positive feedback applied to a switching circuit.
- [ ] To apply RC timing theory to the design of saturating pulse generators and oscillators.
- [ ] To describe the use of an integrated circuit timer as an oscillator.
- [ ] To design an oscillator based on a differential comparator.

The application to switching circuits represents one of the most important and widespread uses of silicon transistors. Unlike an amplifier circuit which usually aims to process signals in an analogue manner without distorting waveshapes, a switching circuit is essentially two-state. The circuit at rest remains in a well defined state until it is switched to the other state, the transition occurring instantaneously in an ideal circuit. This is the basis of digital logic and storage circuits, the essence of the electronic computer.

When reactive components are introduced into switching circuits a timing function is performed. This effect can be used to advantage in the generation of single pulses or a continuous train of pulses.

BJT regions of operation

Consider the simple n-p-n common emitter circuit of Fig. 6.1.

> Leakage currents may be neglected at moderate temperatures.

Off region: If $I_B = 0$ or $V_{BE} \leqslant 0$, then $I_C = 0$. The transistor is in a nonconducting state and is said to be OFF. Since $I_C = 0$, $I_C R_C = 0$ and $V_{out} = V_{CE} = +V_{CC}$.

Active region: If I_B (or V_{BE}) is raised above zero, then I_C rises from zero; $I_C R_C$ increases and $V_{out} = V_{CC} - I_C R_C$ falls.

If $V_{CE} > V_{BE}$, i.e. $V_C > V_B$, and $I_C = 0$ the transistor is said to be **active**. In this state, if I_B (or V_{BE}) is changed slightly then I_C and V_{out} change in sympathy. This is the basis of linear amplification.

> Be careful to note the difference between saturation and the ON state.

> $V_{CE(sat)}$ increases with $I_{C(on)}$, overdrive and temperature.

Saturation region: ON state: If I_B (or V_{BE}) and therefore I_C are increased such that V_C falls to V_B, the transistor is on the verge of **saturation**, defined as the collector-base junction being forward biased.

Further increase of I_B (or V_{BE}), and consequently I_C, leads to V_C falling very close to $V_E(V_{CE(sat)} \approx 100 \text{ mV})$. When V_{CE} cannot fall any further, the transistor is said to be fully saturated or ON.

> Remember that β is subject to wide variability. Just using a typical value for β can lead to incorrect design.

Note that the extra I_B (or V_{BE}) just turns on further the emitter-base junction giving rise to an excess base current, or overdrive, which is used to ensure saturation with a minimum β device. In this condition the β relationship between I_C and I_B no longer holds; the excess I_B adds to I_E without increase of I_C. Also, it is important

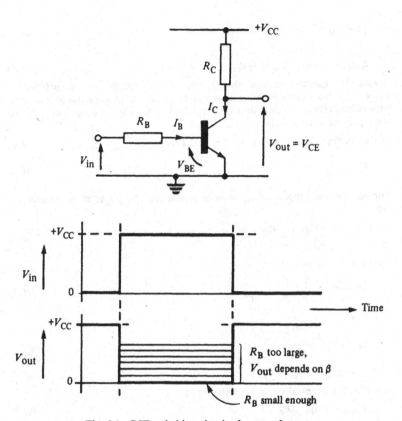

Fig. 6.1   BJT switching circuit plus waveforms.

to recognize that, in practice, owing to the variability of characteristics from device to device and the exponential nature of the input characteristic, a common-emitter transistor is current rather than voltage driven.

Increasing overdrive improves turn-on speed. See Sparkes (1987), Chapter 4.

Simple switching circuits

Single-stage inverter

In switching circuits transistors are driven between the two well defined states, OFF and ON. Passage through the active region is not important except in determining the speed of switching.

Consider the ON and OFF states of a simple switching circuit with current drive as shown in Fig. 6.1.

OFF state:

$$I_B = I_C = 0 \qquad\qquad (6.1)$$

$$V_{in} = V_{BE} \leqslant 0 \qquad \text{(zero or reverse bias)} \tag{6.2}$$

and, since $I_C = 0$,

$$I_C R_C = 0 \quad \text{and} \quad V_{out} = +V_{CC} \tag{6.3}$$

Taking V_{BE} negative ensures that $I_B < 0$ and so speeds up turn-off but if $V_{BE} < -5\,\text{V}$ the emitter-base junction may break down, an undesirable feature to be avoided in timing circuits since the breakdown voltage is not accurately specified for a particular device type.

ON state: Since $V_{out} \approx 0\,\text{V}$, $I_C R_C \approx +V_{CC}$. Therefore $\qquad\qquad$ (6.4)

VCE(sat) may usually be neglected in calculating $I_{C(on)}$.

$$I_{C(on)} = \frac{V_{CC}}{R_C} \tag{6.5}$$

Saturation must be ensured, even for a minimum β device.

The base current supplied, $I_{B(on)}$, must be greater than $I_{C(on)}/\beta$ to ensure saturation. Now

It is usually sufficiently accurate to assume that V_{BE} of a conducting transistor is 0.7 V.

$$I_B = \frac{V_{in} - V_{BE(on)}}{R_B} = \frac{V_{CC} - 0.7}{R_B} \tag{6.6}$$

for $V_{in} = +V_{CC}$, and

$$\frac{V_{CC} - 0.7}{R_B} > \frac{V_{CC}}{R_C} \frac{1}{\beta} \tag{6.7}$$

Therefore

The approximation of Equation 6.9, while not strictly valid for low V_{CC}, can be accommodated by the overdrive.

$$R_B < \beta R_C \frac{V_{CC} - 0.7}{V_{CC}} \tag{6.8}$$

$$< \beta R_C \quad \text{if } V_{CC} \gg 0.7\,\text{V} \tag{6.9}$$

Allowing for a typical overdrive factor of 2,

$$R_{B(max)} = 0.5 \beta_{(min)} R_C \tag{6.10}$$

With reference to Fig. 6.1, note that this circuit inverts; as V_{in} rises V_{out} falls and vice versa. We can summarize this as shown in Table 6.1.

Table 6.1

The OFF and ON states are simply defined by these close approximations. Both states have very low dissipation, P_C.

State	V_{in}	I_C	$V_{out} = V_{CE}$	I_B	$P_C = I_C V_{CE}$
OFF	$\leqslant 0$	0	$+V_{CC}$	0	≈ 0
ON	$+V_{CC}$	$\approx \dfrac{V_{CC}}{R_C}$	≈ 0	$> \dfrac{V_{CC}}{\beta R_C}$	≈ 0

Exercise 6.1 For the circuit of Fig. 6.1, $V_{in} = V_{CC} = +5\,\text{V}$ and $R_C = 1\,\text{k}\Omega$. Calculate R_B such that the transistor is fully saturated with an overdrive factor of at least 2. Assume β lies between 50 and 400.
[*Answer*: $R_B = 22\,\text{k}\Omega$]

This circuit is the basis of an early series of digital integrated circuits, resistor-transistor logic (or RTL).

BJT NOR gate

If a second input is applied via a resistor to the transistor base (Fig. 6.2), the circuit performs a logic function on the two inputs. If either V_1 or V_2 (or both) is taken to

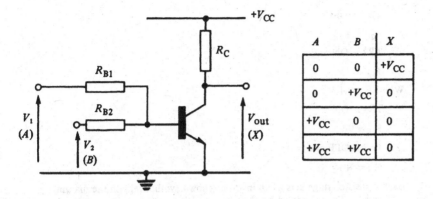

A	B	X
0	0	$+V_{CC}$
0	$+V_{CC}$	0
$+V_{CC}$	0	0
$+V_{CC}$	$+V_{CC}$	0

Fig. 6.2   BJT NOR gate and truth table.

$+V_{CC}$, base current flows, the transistor turns ON and $V_{out} = 0\,V$. This is the logical NOR function, negated (or inverted) OR, which can be expressed by the truth table shown.

Cascaded switching stages

Two or more switching circuits may be cascaded as shown in Fig. 6.3. Assuming that $R_B \gg R_C$ (i.e. no loading), when $V_{in} = 0$,

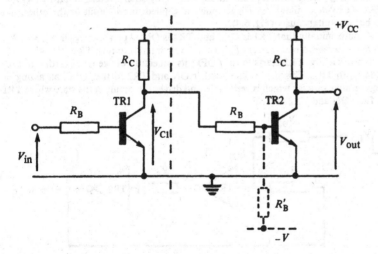

Fig. 6.3   Cascade of two switching stages.

105

TR1 is OFF

$V_{C1} = +V_{CC}$

TR2 is ON and

$V_{out} \approx 0\,V$

When $V_{in} = +V_{CC}$,

TR1 is ON

$V_{C1} \approx 0\,V$

TR2 is OFF and

$V_{out} = +V_{CC}$

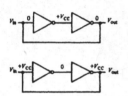

$V_{CE(sat)}$ has a positive temperature coefficient, V_{BE} a negative one.

See Horrocks (1990), Chapter 2.

Each transistor stage acts as an inverter (shown symbolically in the margin).

A resistor R'_B, connected between base and a negative supply, is often included to ensure that the transistor is definitely turned OFF with a reverse biased base-emitter junction. Particularly at high temperatures it is not possible to rely on $V_{CE(sat)}$ being small enough to turn OFF a following transistor.

Positive feedback: the bistable

If the output is connected back to the input (i.e. $V_{in} = V_{out}$) over the two stages, with $V_{in} = 0\,V$ or $V_{in} = +V_{CC}$, the output reinforces the state of the input. This feedback connection is called **positive feedback**. Also, there are two possible states as shown. Both are stable states and the circuit is called a **bistable**. The circuit stays in a stable state unless forced to change by the application of an external signal. This is a 1-bit storage (or memory) cell.

To change the circuit from one stable state to the other, second inputs to TR1 and TR2 are made available. A trigger pulse is applied to one input or the other, not to both simultaneously (Fig. 6.4).

Assume initially that TR1 is OFF and TR2 is ON. Applying trigger $V_{T1} = +V_{CC}$ to R_{B2} generates a base current I_{B2} for TR1 which tends to turn TR1 ON. When V_{C1} falls sufficiently, TR2 starts to turn OFF; its collector voltage rises starting to provide I_{B1} to TR1. This drives TR1 further ON and TR2 further OFF, an aiding or regenerative process which is very rapid and drives the circuit to the state where TR1 is fully ON and TR2 is OFF.

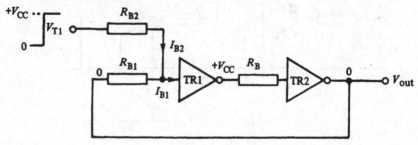

Fig. 6.4   Modification to allow change of state.

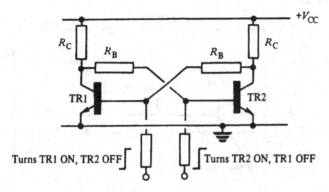

Fig. 6.5   Set–reset (S–R) bistable.

The input signal V_{T1} is now no longer needed and may be removed. Further application of V_{T1} trigger pulses have no effect. To return the circuit to its initial state a similar trigger pulse must be applied to TR2. A complete bistable (or flip-flop) is shown in Fig. 6.5.

There are various methods of triggering, clocking, setting and resetting bistables; they are not described here as they are best treated at logic gate level (e.g. J-K flip-flop). However, it is stressed that a major result of applying positive feedback or regeneration is the very fast change of state.

Stonham (1987, Chapter 4), treats bistables in detail at gate level.

Timing circuits

Timing mechanism

Introducing capacitors into switching circuits can yield controlled delays between changes of state. This is useful in generating pulses of a predetermined duration in response to trigger pulses and can even be used to produce oscillator circuits.

The general mechanism is described with reference to the circuit of Fig. 6.6 for which three conditions are discussed. TR1 is a saturating transistor which can be

An intuitive solution technique for R–C timing circuits is given in Appendix B. It should prove useful in relation to this section.

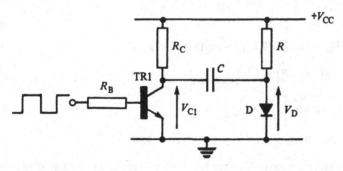

Fig. 6.6   Basic timing circuit.

107

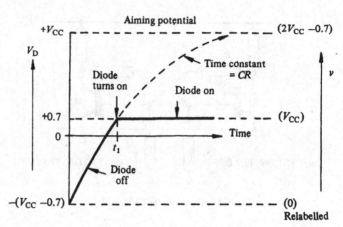

Fig. 6.7 Timing diagram for Fig. 6.6.

driven OFF and ON and may be represented by an open or closed switch respectively.

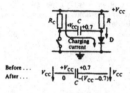

TR1 OFF. With TR1 OFF, C is charged through the R_C and diode path to the state indicated. Diode D is kept conducting by the current through R.

TR1 ON. The switch closes causing V_{C1} to fall from $+V_{CC}$ to 0 V. This negative step of amplitude V_{CC} is applied to the left-hand plate of the capacitor. Since the voltage across a capacitor cannot change instantaneously, the right-hand plate of C is driven negative by the same amount, reverse biasing the diode.

We now have the condition where the right-hand plate of the capacitor charges via R towards $+V_{CC}$ with time constant CR as shown in Fig. 6.7.

When V_D reaches 0.7 V, the diode turns ON again. V_D remains at 0.7 V while the charging current through R is diverted into D.

Let us calculate t_1, the time for which the diode is OFF. Relabelling the voltage axis relative to $-(V_{CC} - 0.7)$ as zero simplifies the calculation.

$$v = (2V_{CC} - 0.7)\,[1 - \exp(-t/CR)] \tag{6.11}$$

Now

$$v = V_{CC} \quad \text{at } t = t_1 \tag{6.12}$$

$$V_{CC} = (2V_{CC} - 0.7)\,[1 - \exp(-t_1/CR)] \tag{6.13}$$

$$\exp(-t_1/CR) = 1 - \frac{V_{CC}}{2V_{CC} - 0.7} \tag{6.14}$$

$$t_1 = CR \ln\left[\frac{2V_{CC} - 0.7}{V_{CC} - 0.7}\right] \tag{6.15}$$

Note that, if $V_{CC} \gg 0.7$ V, $t_1 = CR \ln 2$ or $0.69CR$.

Don't make life unnecessarily difficult! A clearly labelled timing diagram is essential for correct analysis.

It is important to remember this exact relationship as well as the approximation.

Exercise 6.2   Calculate t_1 in terms of CR when $V_{CC} = +5$ V. Is the factor 0.69 a good approximation in this case?

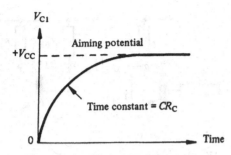

Fig. 6.8 Collector recovery waveform.

TRl turned OFF after t_1. If the switch is opened after t_1 the left-hand plate of the capacitor rises towards $+V_{CC}$ with time constant CR_C (see Fig. 6.8). The charging current flows through R_C and diode D, keeping D ON and V_D constant at 0.7 V. This non-instantaneous rise of TR1 collector voltage (or collector recovery) is expressed by

$$V_{C1} = V_{CC}[1 - \exp(-t/CR_C)] \tag{6.16}$$

Recovery must be nominally complete (allow at least $3CR_C$) before TR1 may be turned ON again. Otherwise C will not have fully charged to the initial conditions assumed in the timing diagram of Fig. 6.7, the negative transient amplitude will be reduced and the expression derived for t_1 (Equation 6.15) must be amended accordingly.

Recovery 90%
    complete after $2.3CR_C$ s
Recovery 99%
    complete after $4.6CR_C$ s
Recovery 99.9%
    complete after $6.9CR_C$ s

Pulse generator

The diode of the previous circuit can be replaced by the emitter-base junction of a transistor, the switch TR2 in Fig. 6.9. Note that the trailing edge of the V_{C2} output pulse is not an abrupt transition, merely an amplified version of the base waveform V_{B2}, when TR2 becomes active. To make this edge sharp, positive feedback must be applied round the circuit.

For the circuit of Fig. 6.9 show that, if $t_1 > T$, the positive V_{C2} output pulse has duration T, independent of the circuit timing characteristic.

Exercise 6.3

Monostable

Positive feedback, or regeneration, can be applied to the simple pulse generator yielding the circuit of Fig. 6.10. When V_{in} goes to $+V_{CC}$, V_{out} does likewise reinforcing the TR1 ON state. This state is unstable; TR2, although turned OFF initially, will turn ON after a time t_1 (unchanged at approximately $0.69CR_{B2}$).

Remember that this approximation is poor for low V_{CC}.

If V_{in} is taken to 0 V before t_1, TR2 will turn ON regeneratively at t_1. However, if V_{in} is kept at $+V_{CC}$ beyond t_1, TR2 will still turn ON after t_1 but, since TR1 is held ON regardless of the state of TR2, the turn ON of TR2 is non-regenerative. In both cases an output pulse of duration t_1 results. (Trigger pulses are usually derived by a CR differentiator to make them short and sharp.)

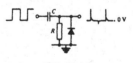

Since this circuit has one stable state and one unstable (or quasistable) it is called a **monostable** circuit.

Also called a **one-shot**.

109

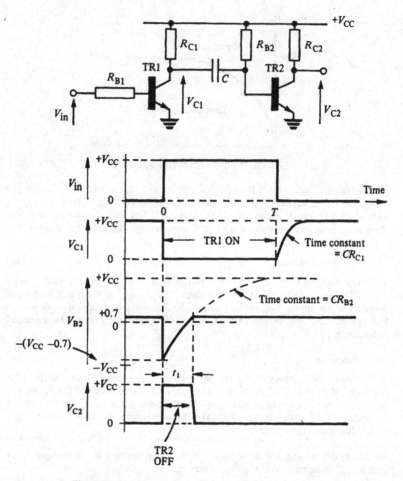

Fig. 6.9 Pulse generator with waveforms for $T > t_1$.

It must be noted that breakdown of the base-emitter junction of TR2 could occur if V_{CC} were greater than approximately 5.7 V ($|BV_{BE}|$ usually lies between 5 and 8 V for a silicon transistor). The negative excursion of V_{B2} must be limited to less than 5 V by restricting the magnitude of V_{CC} or by introducing special circuitry, otherwise the timing t_1 may be affected.

Design Example 6.1 Design a monostable circuit to generate positive-going output pulses of 5 V amplitude and 100 μs duration. For the transistors you may assume: $I_{C(on)} = 5$ mA, $V_{BE(on)} = 0.7$ V, $V_{CE(sat)} = 0$ V, $|BV_{BE}| \geqslant 6$ V and $\beta \geqslant 40$.

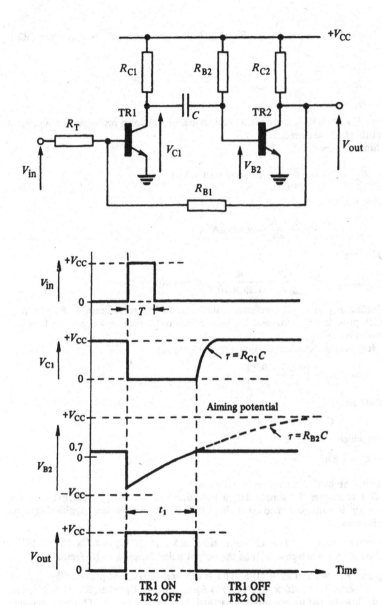

Fig. 6.10 Monostable circuit with waveforms.

Solution: Using the basic circuit of Fig. 6.10 and assuming $R_B \gg R_{C2}$: when TR2 is OFF,

$$V_{out} = +V_{CC}$$

Therefore

$$V_{CC} = +5\,V$$

Since $V_{CC} < |BV_{BE}|$, there are no emitter-base breakdown problems and no circuit modifications are needed.

Since $I_{C(on)} = 5\,mA$,

$$R_{C1} = R_{C2} = \frac{V_{CC}}{I_{C(on)}} = 1\,k\Omega \text{ (preferred value)}$$

For minimum β,

$$I_{B(on)} = \frac{I_{C(on)}}{\beta} = \frac{5 \times 10^{-3}}{40} = 125\,\mu A$$

Therefore

$$R_{B2(max)} = \frac{V_{CC} - 0.7}{I_{B(on)}} = \frac{4.3}{1.25 \times 10^{-4}} = 34.4\,k\Omega$$

Selecting R_{B2} as $27\,k\Omega$ (preferred value) should ensure saturation. R_T and R_{B1}, which provide triggering and feedback respectively, may be assumed to have the same value as R_{B2}.

Now consider the timing; Equation 6.15 applies

$$t_1 = CR_{B2}\ln\left[\frac{2V_{CC} - 0.7}{V_{CC} - 0.7}\right]$$

Substituting,

$$10^{-4} = C \times 2.7 \times 10^4 \times 0.77$$

from which

$$C = 4.8\,nF$$

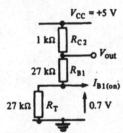

$V_{CC} = +5\,V$

$1\,k\Omega$ R_{C2}

V_{out}

$27\,k\Omega$ R_{B1}

$I_{B1(on)}$

$27\,k\Omega$ R_T $0.7\,V$

(Use 4.7 nF as the nearest preferred value.)

This completes the simple design procedure but we must check that collector recovery is complete and that R_{B1} and R_T do not affect significantly the performance.

Collector recovery: The collector time constant ($CR_{C1} = 4.7 \times 10^{-9} \times 10^3 = 4.7\,\mu s$) is very much shorter than the output pulse duration – satisfactory.

R_{B1} *and* R_T: When TR2 is OFF, TR1 is ON and the current provided by R_{C2} and R_{B1} in series is $4.3/28 \times 10^3$, i.e. $153.6\,\mu A$. Of this R_T takes $0.7/27 \times 10^3$, i.e. $25.9\,\mu A$ when the trigger input is earthed, leaving $127.7\,\mu A$ as TR1 base current. Ignoring component tolerances, this is just sufficient to ensure that TR1 is saturated.

The current taken by R_{B1} does affect the output voltage level reducing it to $5 - 4.3/(28 \times 10^3)$ or $4.85\,V$.

Note that output voltage levels and pulse duration depend on the tolerances of

C, R_{B1}, R_{B2}, R_{C2}, $V_{CE(sat)}$ and V_{CC}. It would be possible to meet the specification exactly by making V_{CC} and R_{B2} adjustable, V_{CC} for amplitude and R_{B2} for timing. The component values which approximately meet the specification are

$$V_{CC} = +5\,V$$

$$R_{C1} = R_{C2} = 1\,k\Omega$$

$$R_{B1} = R_{B2} = R_T = 27\,k\Omega$$

$$C = 4.7\,nF$$

Astable

In order to produce continuous oscillation from two-state switching circuits, both states must be unstable; the circuit is then said to be **astable**. If two timing circuits are connected in cascade and the output connected to the input, one unstable state (duration t_1) is followed by a second (duration t_2), then t_1, t_2 and so on. The frequency of oscillation (f_{osc}) is given by

$$f_{osc} = \frac{1}{\text{Period}} = \frac{1}{t_1 + t_2} \tag{6.17}$$

and the output waveform is a squarewave if $t_1 = t_2$.

In the circuit of Fig. 6.11 collector transients are cross-coupled to the opposite transistor base, e.g. when TR2 turns ON, TR1 is turned OFF for a time t_1 ($\approx 0.69 C_1 R_{B1}$). At the end of t_1, TR1 turns ON turning TR2 OFF for a time t_2 ($\approx 0.69 C_2 R_{B2}$). This cycle repeats.

Note that each change of state is regenerative giving sharp falling edges to both collector waveforms but the rising edges suffer from collector recovery rounding. Collector recovery must be complete and emitter-base breakdown avoided to retain predictable timing.

Special techniques may be introduced to improve the collector waveforms and to allow high-voltage swings to be accommodated without timing problems introduced by emitter-base breakdown.

Equation 6.15 gives the accurate timing if collector recovery is complete and emitter-base breakdown does not occur.

Design a cross-coupled astable to give a 5 V peak-to-peak squarewave output at 1 kHz. Assume the data given in Design Example 6.1.

Design Example 6.2

Solution: Since the output is a squarewave, the circuit is symmetrical, i.e. $C_1 = C_2$, $R_{B1} = R_{B2}$ and $R_{C1} = R_{C2}$. Assuming $V_{CE(sat)} = 0\,V$ and that collector recovery is complete,

$$V_{out(peak-to-peak)} = +5\,V$$

Therefore

$$V_{CC} = +5\,V$$

For $I_{C(on)} = 5\,mA$ and $V_{CC} = +5\,V$, $R_C = 1\,k\Omega$. Also, $R_B = 27\,k\Omega$ (same calculation as in Design Example 6.1). To calculate C, since $f_{osc} = 1\,kHz$,

$$t_1 = t_2 = t = 500\,\mu s$$

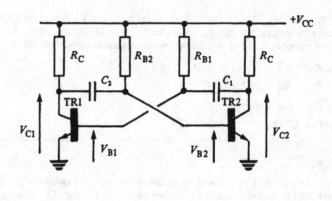

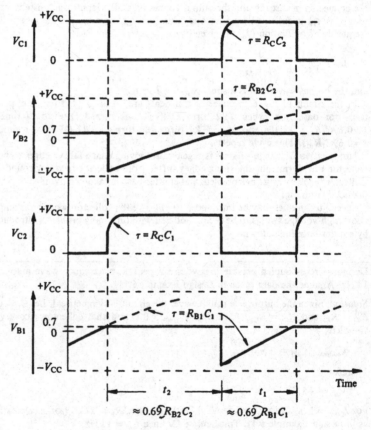

Fig. 6.11   Cross-coupled astable circuit plus waveforms.

Now

$$t = CR_B \ln \left[\frac{2V_{CC} - 0.7}{V_{CC} - 0.7} \right]$$

i.e.

$$5 \times 10^{-4} = C \times 27 \times 10^3 \times 0.77$$

Therefore

$$C = 24\,\mathrm{nF}$$

(Use 22 nF — nearest preferred value.) Checking that collector recovery is complete:

$$CR_C = 24 \times 10^{-9} \times 10^3 = 24\,\mu s$$

and the half-period t is therefore approximately 20 collector time constants, sufficient for effectively complete recovery, and no modification to the timing diagram or equation is necessary. Also, since $V_{CC} < |BV_{BE}|$, there is no danger of emitter-base breakdown.

Collector waveform improvement

In cross-coupled monostable and astable circuits, when a transistor whose collector is connected to a timing capacitor turns OFF, the rising edge of the collector voltage waveform is a $(1 - \exp)$ with time constant CR_C (Fig. 6.8). This rounded edge can be improved by one of several techniques including the use of an isolation diode or an emitter follower.

Isolating diode

Two extra components, a diode D and a resistor R_X are introduced as shown in Fig. 6.12. When TR2 goes OFF, TR1 turns ON and V_{BE1} is held at 0.7 V. Point X charges towards $+V_{CC}$ on a $(1 - \exp)$ waveform with time constant CR_X. However, the diode is reverse biased and V_{C2} can rise rapidly to $+V_{CC}$ isolated from the recovery by the diode, hence the term **isolating diode**.

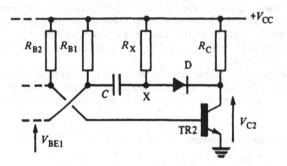

Fig. 6.12 Use of an isolating diode.

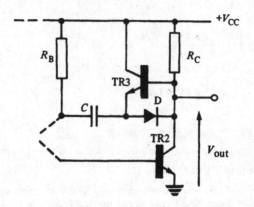

Fig. 6.13   Edge improvement using an emitter-follower.

It is important to note that, when TR2 is ON, its collector load is $R_C\|R_X$. Therefore R_{B2} must be chosen to supply sufficient base current in order to ensure that TR2 is driven fully ON with this modified load.

Also, timing is affected by the introduction of the isolating diode. Now, at turn-ON, the collector voltage (V_{C2}) must fall by approximately 0.7V before the isolating diode conducts and couples a ($V_{CC} - 0.7$) negative-going transient via C to the base of TR1 (compared with a transient of amplitude V_{CC} in the unmodified circuit). A new timing diagram must be constructed and a corresponding timing equation calculated. The recovery waveform, even though it has moved from the collector to point X, must still be complete for accurate timing prediction.

If both collector waveforms of a cross-coupled astable circuit are required to be improved then isolating diodes can be incorporated in both halves of the circuit.

Emitter follower

Addition of an emitter-follower transistor, TR3 in Fig. 6.13, provides a current gain of $(1 + \beta)$ between its base and emitter, reducing the collector recovery time constant from CR_C to $CR_C/(1 + \beta)$.

The positive-going edge of the V_{C2} waveform (TR2 turning OFF) turns ON TR3 and the current gain from TR3 is realized. However, on the negative-going edge, TR3 is turned OFF since its emitter is held by the capacitive load. A **pull-down** diode D, connected so as to conduct when TR3 is OFF, is added to transmit this edge via C to the opposite base.

The circuit timing is again affected by these modifications since C does not charge fully to the $+V_{CC}$ supply; also, the negative transient amplitude transmitted through C is reduced to ($V_{CC} - 0.7$) owing to the voltage drop across the diode.

Junction breakdown protection

Junction breakdown

If the reverse bias applied to a diode is increased beyond a certain limit, the junction

116

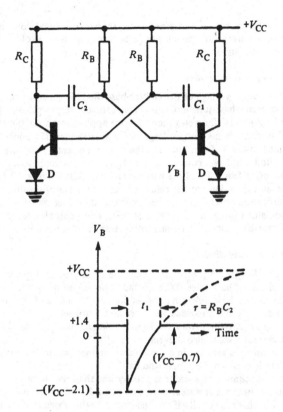

Fig. 6.14 Emitter-base protection and modified timing diagram.

will break down (at a voltage termed BV). This applies to the emitter-base and collector-base junctions of a BJT and there is a similar limit to V_{CE}. At breakdown the diode voltage is nominally fixed. In all of the timing circuits already described, problems can be introduced if the supply voltage is such that the emitter-base and/or the collector-base junction breaks down on a negative-going transient.

Emitter-base junction. If the expected negative base transient to $-(V_{CC} - 0.7)$ is greater than $|BV_{BE}|$ then the base excursion is limited to the breakdown voltage which varies from one device to the next. Since the starting point of the $(1 - \exp)$ waveform is in error the timing is affected and unpredictable (although it is shorter than expected).

For low-power silicon planar transistors, $|BV_{BE}|$ is *usually* specified as being not less than 6 V.

Collector-base junction. If the collector-base voltage rating is exceeded by the negative-going base transient, the collector voltage is forced to fall. This is capacitively coupled to the opposite base prematurely turning OFF the transistor which is supposed to be ON. Timing can be violently disturbed.

As well as the above timing effects it is important to appreciate that if a diode in breakdown is forced to pass a large current in discharging a timing capacitor then the device may be destroyed. As a general rule, therefore, if breakdown of junctions is approached the protection circuitry must be incorporated to avoid failure and to permit accurate timing design.

Protection using emitter diodes

Since the breakdown voltage of the emitter-base diode is generally much less than that of the collector-base junction, it is often sufficient to protect only the former. The circuit of Fig. 6.14 illustrates the technique applied to an astable circuit.

Diodes (D) with a breakdown voltage rating greater than V_{CC} allow the emitter-base junctions to break down but limit the reverse current to the safe level of the protection diode leakage current. Since $V_{B(ON)} = 1.4\,\text{V}$ and $V_{C(ON)} = 0.7\,\text{V}$ the timing is modified from that of the basic unprotected circuit. It is left to the reader to calculate an expression for the delay time (t_1). One major disadvantage of this circuit modification arises from the conducting diode voltage drop at the emitter of the ON transistor raising the ON collector voltage to approximately 0.7 V.

The collector-base junctions remain unprotected in this circuit.

Protection using base diodes

In the circuit of Fig. 6.15 protection diodes are connected in series with the bases of the transistors. The diodes (D) have the same action as in the previous case; allowing the emitter-base junction to break down but limiting the reverse current to the diode leakage level. Provided that the diodes are selected with a breakdown voltage rating greater than twice the supply voltage, collector-base breakdown at significant current level is also prevented.

Again the timing is affected in a predictable manner but, in this case, the output voltage swing is between the supply and earth.

Since the capacitance of a diode is usually less than the input capacitance of a BJT, when the negative transient occurs at point X the diode turns OFF quickly leaving the transistor to turn itself OFF (slowly!). In this absence of turn-OFF drive, resistors R' are usually included to improve the turn-OFF time.

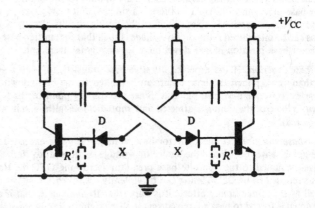

Fig. 6.15   Base diode protection.

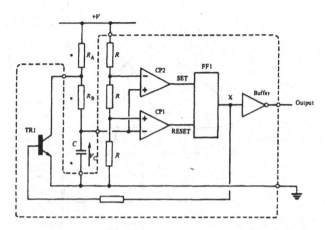

Fig. 6.16 555 timer configured for astable operation. (*external components.)

An integrated circuit timer

A very useful 8-pin integrated circuit (made by several manufacturers as NE555, LM555 etc.) provides monostable or astable operation with a minimum of external components, e.g. the astable configuration requires only one capacitor and two resistors.

Clayton (1975, Chapter 3) provides further information on the 555 and other integrated timers.

The full circuit of the 555 timer is rather complex but is worth considering in the block diagram form of Fig. 6.16. It consists essentially of a resistive bias chain, two comparators, a set–reset bistable, a switching transistor and an inverting output buffer.

A bias chain consisting of three equal-valued resistors (R) makes use of a very important property of integrated circuit components – while it is very difficult to control the manufacturing process to provide resistor tolerances better than 10%, resistor ratios can be determined to within 1%. If components are in close physical proximity (and thus subject to identical processing conditions), their relative parameters are determined by their dimensions and they track with temperature over a wide operating range. The three equal resistors in the timer circuit generate reference voltages $2V/3$ and $V/3$ from the supply voltage (V) irrespective of the supply voltage, temperature and absolute value of the resistors. We shall see later the significance of this design feature.

In astable operation, external resistors R_A and R_B charge and discharge the external capacitor C conditional upon the state of the internal switching transistor (TR1). Two comparators (CP1 and CP2) compare the capacitor voltage with the bias chain references and provide SET and RESET inputs to a bistable or flip-flop (FF1) whose output (X) in turn determines whether TR1 is ON or OFF. The output of the bistable is buffered to give a low impedance circuit output.

Operation of the circuit is best described from the switch-on state with C discharged ($V_C = 0$).

In practice, at a temperature of 25 °C and for a load current of 5 mA, the output voltage is (V – 1.4) in its high state and 0.1 V in its low state.

(a) When the supply voltage ($+V$) is first applied, since $V_C = 0$, the inverting comparator (CP1) applies a logic 1 RESET level to the bistable while the SET output of the non-inverting comparator (CP2) is at logic 0. The output X of

119

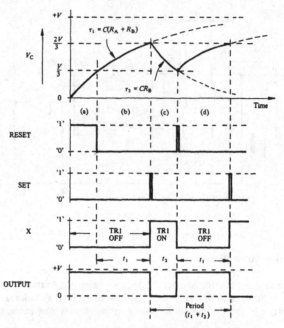

Fig. 6.17   Waveforms for 555 astable.

the bistable in its RESET state is at logic 0, holding TR1 OFF and, inverted by the buffer, gives an output voltage of approximately $+V$. The capacitor charges towards the aiming potential of $+V$ with a time constant $C(R_A + R_B)$ (see Fig. 6.17a).

(b) When V_C reaches the $V/3$ reference for CP1, the RESET signal is removed and C continues charging while the SET signal and X remain at logic 0 and the output stays at $+V$ (Fig. 6.17b).

(c) When V_C reaches the $2V/3$ reference for CP2, the bistable is SET, X switches to logic 1 and turns TR1 ON. C now discharges (immediately removing the SET signal) towards earth with a time constant CR_B. The circuit output is now at approximately 0 V (Fig. 6.17c).

(d) When C has discharged to $V/3$ the RESET signal is applied and X switches to logic 0 turning TR1 OFF. The circuit output switches to $+V$ (Fig. 6.17d). The cycle of charging and discharging C repeats giving a rectangular output waveform whose part periods (t_1 and t_2) are now calculated.

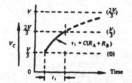

To calculate t_1 consider the charging period (d) in Fig. 6.17.

$$V_C = V[1 - \exp(-t/\tau_1)]$$

where $\tau_1 = C(R_A + R_B)$. Relabelling the voltage axis as shown gives

$$\frac{V}{3} = \frac{2V}{3}[1 - \exp(-t_1/\tau_1)]$$

from which

$$t_1 = \ln 2 [C(R_A + R_B)]$$
$$= 0.693 C(R_A + R_B) \qquad (6.18)$$

For the discharge period (c) in Fig. 6.17

$$V_C = \frac{2V}{3} \exp(-t/\tau_2)$$

where $\tau_2 = CR_B$. Therefore

$$\frac{V}{3} = \frac{2V}{3} \exp(-t_2/\tau_2)$$

and

$$t_2 = 0.693 CR_B \qquad (6.19)$$

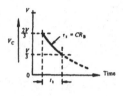

The frequency of oscillation (f) is given by

$$f = \frac{1}{t_1 + t_2} = \frac{1}{0.693 C(R_A + 2R_B)} \qquad (6.20)$$

and is theoretically independent of the supply voltage (V) owing to the bias chain providing comparator reference voltages proportional to V. In practice the astable frequency exhibits a supply voltage dependence of only 0.1% per volt and is stable to 0.005% per °C. This superb performance plus the requirement for only one timing capacitor makes the 555 integrated timer a very useful device for generating rectangular waveforms up to 200 kHz.

There are a host of other applications, including monostable operation, for which the 555 timer is suited; the reader is advised to consult manufacturers' literature for further information.

Using a 555 integrated circuit timer (Fig. 6.16), design an oscillator to produce a **Design Example 6.3** 10 V peak-to-peak rectangular waveform at a frequency of 50 kHz. The mark-to-space ratio of the output may be freely chosen and the output may be assumed to be lightly loaded.

Solution. Since the required output swing is 10 V peak-to-peak and assuming a load current of 5 mA, the supply voltage required is

$$V = 0.1 + 10 + 1.4 = 11.5 \text{ V}$$

Equation 6.20 gives the oscillation frequency in terms of the external components C, R_A and R_B

$$f = \frac{1}{0.693 C(R_A + 2R_B)} = 50 \text{ kHz}$$

Therefore

$$C(R_A + 2R_B) = \frac{1}{0.693 \times 5 \times 10^4} = 28.86 \, \mu s$$

Now, applications data for the 555 timer imply that C should be limited to the range 1 nF to 100 μF and that $(R_A + 2R_B)$ should lie between 1 kΩ and 10 MΩ.

With C selected as 10 nF, $(R_A + 2R_B) = 2.886$ kΩ, which may be realized by choosing

$$R_A = 1.8\,\text{k}\Omega$$

and

$$R_B = 560\,\Omega$$

Checking the results,

$$f = \frac{1}{0.693 \times 10^{-8} \times 2.92 \times 10^3}$$

$$= 49.4\,\text{kHz}$$

which is satisfactory. We can also calculate the output mark-to-space ratio as

$$\frac{t_1}{t_2} = \frac{R_A + R_B}{R_B} = \frac{2.36}{0.56} = 4.2$$

Differential comparator oscillator

A single differential comparator plus several resistors and a capacitor can be used as an oscillator whose output is directly compatible with logic circuits. Consider the circuit of Fig. 6.18a which shows a comparator with positive feedback applied (R_1 and R_2) and a further resistor R and capacitor C. Upper and lower threshold levels (UTL and LTL) are established by the feedback as described in Chapter 5. If the output voltage (V_{out}) has high and low logic levels designated V_{OH} and V_{OL} respectively, then

$$UTL = kV_{OH}$$

and

$$LTL = kV_{OL}$$

where $k = R_1 / (R_1 + R_2)$.

Assuming that initially the capacitor voltage (V_C) is negative with respect to the LTL, $V_{out} = V_{OH}$, V_X at the non-inverting input of the comparator is at the UTL and the capacitor charges through R towards V_{OH}. When V_C reaches the UTL the comparator changes state, V_{out} switches to V_{OL} and V_X to the LTL. The capacitor now discharges towards V_{OL} until V_C reaches the LTL when the circuit again changes state. This cycle continues with the timing dictated by the charging and discharging of the capacitor.

For this circuit to operate, V_{OH} must be positive and V_{OL} negative. Why?

We can analyse the waveform of V_C (Fig. 6.18b) and determine the frequency of oscillation.

During the part period t_1, the capacitor is discharging from the UTL towards the aiming potential V_{OL} with a time constant CR. By considering the exponential decay, it can be shown that

$$t_1 = CR \ln \left[\frac{kV_{OH} - V_{OL}}{(k-1)V_{OL}} \right] \tag{6.21}$$

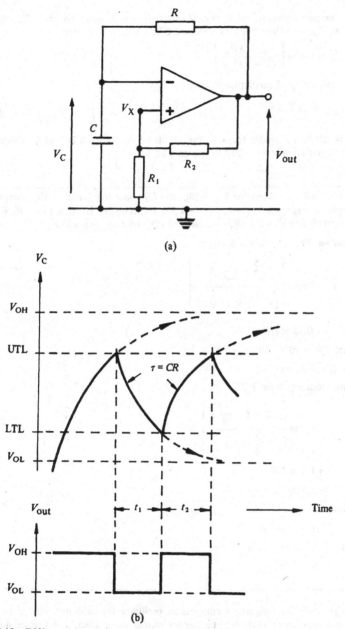

Fig. 6.18 Differential comparator oscillator (a) with its timing and output waveforms (b).

For the part period t_2, the V_C waveform is a $[1 - \exp]$ starting at the LTL and aiming at V_{OH} with the same time constant CR; t_2 can be evaluated as

$$t_2 = CR \ln \left[\frac{V_{OH} - kV_{OL}}{(1 - k)V_{OH}} \right] \tag{6.22}$$

The frequency of oscillation f_{osc} is given by

$$f_{osc} = \frac{1}{t_1 + t_2} \tag{6.23}$$

(It is left to the reader to prove the results of Equations 6.21 and 6.22 using the techniques already described in this chapter.)

Worked Example 6.1

In the oscillator circuit of Fig. 6.18a, the high and low limits of the comparator output voltage are $+3.5\,V$ and $-0.5\,V$ respectively. If $R_1 = 1\,k\Omega$, $R_2 = 3\,k\Omega$, $R = 10\,k\Omega$ and $C = 0.1\,\mu F$, calculate the frequency of oscillation.

First calculate t_1 and t_2.

Solution. From Equation 6.21,

$$t_1 = CR \ln \left[\frac{\dfrac{3.5}{4} - (-0.5)}{\left(-\dfrac{3}{4}\right) \times (-0.5)} \right]$$

$$= 0.32 \times CR$$

Since $CR = 10^4 \times 10^{-7}\,s = 10^{-3}\,s$,

$$t_1 = 0.32\,ms$$

Also, from Equation 6.22,

$$t_2 = CR \ln \left[\frac{3.5 - \left(-\dfrac{0.5}{4}\right)}{\dfrac{3}{4} \times 3.5} \right]$$

$$= 1.3 \times CR$$

$$= 1.3\,ms$$

Therefore

Now the total period and hence the oscillation frequency.

$$t_1 + t_2 = 1.62\,ms$$

and

$$f_{osc} = 617\,Hz$$

Design Note: In designing a comparator oscillator the capacitor voltage swing is made a substantial fraction of the output voltage swing since, if too small, noise in the circuit causes significant timing variations. Also, if the feedback ratio (k) approaches unity, the exponential timing waveforms pass relatively slowly through the threshold levels, again causing timing variations. In practice, values of k ranging from 0.1 to 0.5 prove satisfactory.

This circuit configuration is not restricted to differential comparators; operational amplifiers also may be used.

Summary

The saturating bipolar transistor with its ability to be driven between two well defined levels is the basis for simple digital logic circuits such as the inverter and the NOR gate. Design of such circuits is straightforward – the values of the resistors which define the collector current of an ON transistor and guarantee the necessary base current drive are easily calculated. By applying positive feedback around two cascaded inverters we have also achieved a storage function in the bistable circuit which, by virtue of the feedback, can be switched very rapidly between its two stable states.

Timing circuits, such as the monostable and astable, rely on a capacitive coupling between switching stages introducing an unstable state, the duration of which is governed by the rate of charging of the capacitor. While the calculation of the timing relationships appears to be complex, methodical translation of the circuit operation into a timing diagram is vitally important for accurate analysis. Also, a circuit designer must recognize practical problems such as collector recovery and junction breakdown which can detract from the precision of timing circuits and, if necessary, introduce modifications to minimize their effects. We have also seen the application of integrated circuits to realize timing functions.

It must be recognized that the approaches considered in this chapter serve only as an introduction and are not the only ones in widespread use. Indeed, non-saturating current-steering switching circuits (based on differential amplifiers) are much faster than those involving saturation. In both logic and timing systems there are a plethora of techniques including, of course, the use of dedicated integrated circuits. However, fundamentals are important; the fact remains that most timing circuits are based on the charging and discharging of capacitors.

Problems

6.1 Design a single transistor switching stage to the following specification:
   Supply voltage $= +10\,\text{V}$
   $V_{\text{in(OFF)}} = 0\,\text{V}$
   $V_{\text{in(ON)}} = +5\,\text{V}$
   Transistor $I_{\text{C(ON)}} \approx 2\,\text{mA}$
   Transistor $\beta \geqslant 50$

6.2 What is the logic function of the circuit shown in Fig. 6.19? Assume that the transistor βs are greater than 100 and that $V_{\text{in(1)}}$ and $V_{\text{in(2)}}$ can each have, independently, either of the states $0\,\text{V}$ and $+10\,\text{V}$.

6.3 With reference to the circuit of Fig. 6.20, determine the waveform for V_{C3} resulting from the input pulse shown. Assume that all transistors have $\beta \geqslant 100$.

6.4 In the monostable circuit of Fig. 6.10, $V_{\text{CC}} = +5\,\text{V}$, $R_{\text{C1}} = R_{\text{C2}} = 4.7\,\text{k}\Omega$,

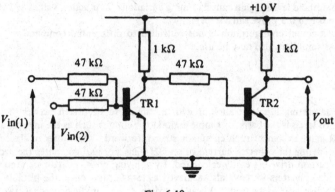

Fig. 6.19

$R_T = R_{B1} = R_{B2} = 47 \text{k}\Omega$ and $C = 10 \text{nF}$. Assuming that the minimum β of the transistors is 50, determine the duration of the output pulse in response to a very short input trigger pulse of amplitude $+5 \text{V}$.

6.5 Using the circuit configuration of Fig. 6.10, design a monostable circuit to produce an output pulse of 4.5 V amplitude and 1 ms duration in response to a very short input trigger pulse. Assume that the minimum β of the transistors is 50 and that each transistor conducts with a collector current of approximately 2 mA.

6.6 In the astable circuit of Fig. 6.11, $V_{CC} = +5 \text{V}$, $R_C = 1 \text{k}\Omega$, $R_{B1} = R_{B2} = 10 \text{k}\Omega$, $C_1 = 10 \text{nF}$ and $C_2 = 47 \text{nF}$.

Determine the frequency of oscillation and the duration of the positive-going output pulses at the collector of TR2. Assume that the minimum β of the transistors is 50.

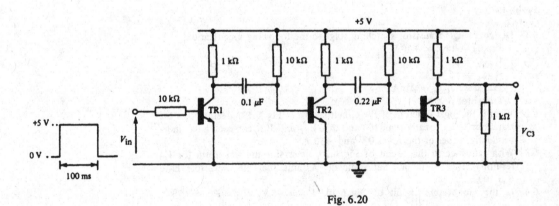

Fig. 6.20

6.7 The base-emitter voltage of a conducting BJT depends on temperature. For the astable circuit of Problem 6.6, determine whether the output frequency rises or falls as a result of an increase in temperature.

6.8 Using the circuit configuration of Fig. 6.11, design an astable circuit to give a 4 V peak-to-peak squarewave at a frequency of approximately 5 kHz. Assume that the minimum β of the transistors is 50 and that each transistor conducts with a collector current of approximately 4 mA.

6.9 Design a cross-coupled astable multivibrator to meet the following specification:

    pulse repetition frequency $\approx$ 5 kHz

    output mark-to-space ratio $\approx$ 4:1

    peak-to-peak output voltage $\approx$ 10 V

When a transistor is ON, its collector current should be approximately 5 mA. It may be assumed that the transistors have a minimum β of 50, $|BV_{CB}| \geqslant 30$ V and $|BV_{BE}| \geqslant 5$ V; also that a forward biased silicon junction exhibits a voltage drop of 0.7 V. It is not necessary to include circuitry to enhance the rise time of the collector waveform but collector recovery time should be considered.

6.10 Using a 555 integrated circuit timer, design an oscillator to produce a train of positive pulses of 570 μs duration at a frequency of 1.25 kHz.

6.11 Using the circuit of Fig. 6.18a, design a comparator oscillator to give an output frequency of approximately 10 kHz. The comparator output logic levels are $V_{OH} = +5$ V and $V_{OL} = -5$ V and the peak-to-peak capacitor voltage should be approximately 2.5 V. What is the mark-to-space ratio of the output voltage?

6.12 A cross-coupled astable multivibrator incorporates emitter diodes for junction protection. Assuming that a forward biased junction exhibits a voltage drop of 0.7 V, derive an expression for the transistor OFF-time in terms of timing components (C and R_B) and supply voltage (V_{CC}).

6.13 Repeat Problem 6.12 with the emitter diodes replaced by base diodes.

7 Field-effect transistors and circuits

□ To explain the structure and operation of both JFETs and MOSTs plus the characteristics and equations governing their behaviour.

□ To explain the difference between n-channel and p-channel devices, and between depletion and enhancement FETs.

□ To construct a small-signal a.c. model for a FET.

□ To bias depletion and enhancement FETs.

□ To explain the properties of FET common-source, series feedback and source follower amplifiers.

□ To use a FET as a voltage-variable resistor.

□ To use FETs as series and shunt switches and to explain the gate drive requirements for the ON and OFF states.

□ To describe the way in which MOS logic circuits are designed.

□ To describe the operation of CMOS gates.

As early as 1925, Lilienfield discovered that the conductivity of a material could be varied by applying an electric field perpendicular to the current flow – the fundamental principle of the field effect. However, it was not until some 30 years later that the first successful field-effect transistors (FETs) were made.

A FET has three terminals: a **source**, a **drain** and a **gate**. The semiconductor region between source and drain is called a **channel** and its conductivity is controlled by the potential of the gate terminal.

There exist two types of FET, the **junction-gate device** (JUGFET or JFET), in which the gate is separated from the channel by a (normally reverse biased) p–n junction, and the **insulated-gate FET** (IGFET) in which a thin layer, usually of silicon oxide, provides isolation of the gate electrode from the channel. Because of its structure, this latter device is also termed the **metal-oxide-semiconductor-transistor** (MOSFET or MOST). In this text the acronyms JFET and MOST are used to distinguish between the two types of field-effect transistor.

Both JFETs and MOSTs exhibit an extremely high input resistance which accounts for their popularity in amplifier circuits and the low power requirements of logic circuitry using MOSTs, together with very small device dimensions, make possible the very large-scale integrated circuits (VLSI) of today.

The JFET

Structure and operation

General references on FETs are Sparkes (1987), Chapters 1, 3 and 5, and Millman and Grabel (1987), Chapter 4.

The structure of a junction field-effect transistor (JFET) is shown in schematic form in Fig. 7.1a. For an n-channel device the source and drain terminals are at

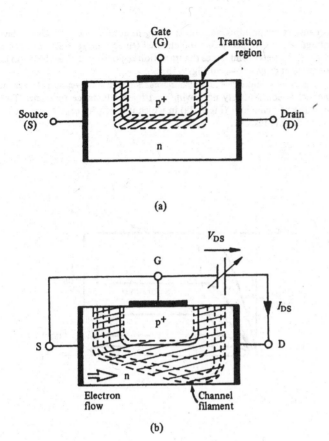

(a)

(b)

Fig. 7.1   n-channel JFET structure: (a) in equilibrium, (b) with $V_{GS} = 0$ V and increasing V_{DS} applied.

opposite ends of an n-type silicon region into which is diffused a p^+ gate region. In equilibrium, with no connection made to any terminal, the transition region associated with the p–n junction extends partly into the n-type channel between source and drain (also into the p^+ gate region) but not to the extent of blocking the channel. The transition region is depleted of mobile carriers and may be regarded as an insulator serving to delineate the conductive channel boundary.

Let us determine the shape of the output and transfer characteristics for an n-channel JFET.

If the gate is connected to the source ($V_{GS} = 0$ V) and if a small potential (V_{DS}) is applied between drain and source (drain positive with respect to source), current flows through the channel in the form of electron conduction from source to drain since the channel is n-type. However, the drain current (I_{DS}) is defined in the conventional direction, that of positively charged current carriers. Small variations

In a FET, only one type of current carrier (the majority carrier) is involved in the conduction process. For this reason, FETs are called **unipolar** transistors (c.f. bipolar junction transistors).

129

of V_{DS} cause corresponding variations of I_{DS} in accordance with Ohm's law. Since the voltage V_{DS} is dropped along the channel length, the reverse bias between gate and channel is greater (and hence the transition region is wider) at the drain end than at the source end, as shown in Fig. 7.1b.

As V_{DS} is steadily increased, I_{DS} also increases but at a slower and slower rate since the channel is continuously narrowing and its conductance reducing. This is the **ohmic region** of behaviour (Fig. 7.2a) in which it is helpful to think of the device as a voltage-variable resistor.

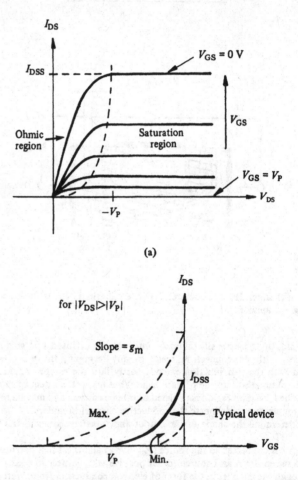

(a)

Fig. 7.2 n-channel JFET characteristics: (a) idealized output characteristic, (b) transfer characteristic. (Note: V_p is negative.)

It might be expected that, when V_{DS} is increased to a critical level, the transition region widens sufficiently to completely block the channel and stop conduction. In fact the channel reduces to an extremely narrow filament and the electron flow is limited to a virtually constant rate. The channel becomes saturated (Fig. 7.1b). At this level, the drain current is termed the drain-source saturation current (I_{DSS}). If V_{DS} is increased further, the length of the channel filament increases and, ideally, I_{DS} remains constant at I_{DSS}. This is the **saturation region** of operation (Fig. 7.2a).

The electric field due to V_{DS} is in the direction source to drain.

One curve on the output characteristic has now been traced. Curves for other values of V_{GS}, the input control parameter, are also shown on Fig. 7.2a. For an n-channel device V_{GS} must be less than or equal to zero otherwise the gate is forward biased and consequent gate current destroys the inherent high input resistance of the device. When V_{GS} is sufficiently negative and the reverse bias is of such a magnitude that the transition region extends completely across the channel, no current other than a very small leakage current can flow between source and drain. At this gate voltage, the channel is said to be **pinched-off** and $V_{GS} = V_p$, the gate pinch-off voltage. This value of gate-source voltage (V_p) has the same magnitude as the drain-source voltage at which the channel saturates with $V_{GS} = 0\,V$. Figure 7.2a has been annotated accordingly. In practice, the output characteristic departs from the idealized version in that the curves have non-zero slope in the saturation region thus exhibiting a finite output resistance.

In correspondence with thermionic valve terminology, the ohmic and saturation regions are referred to as **triode** and **pentode** regions respectively.

Saturation is often called pinch-off but in this text, to avoid possible confusion between these terms, their single definitions are used.

For an n-channel JFET, V_{GS} and V_p are both negative.

How a JFET can be used as a voltage-variable resistor or switch in the ohmic region and as an amplifier in the saturation region is described later in this chapter.

The transfer characteristic of a JFET in its saturation region can be derived by translating the output characteristic onto I_{DS}, V_{GS} axes (current output, voltage input) as shown in Fig. 7.2b. The slope of the transfer characteristic defines the transconductance or mutual conductance (g_m or g_{fs}). A major disadvantage of the JFET is the variability (or spread) within a device type of V_p, I_{DSS} and g_m, limits of which are illustrated in Fig. 7.2b together with the typical transfer characteristic.

For a typical n-channel JFET,

I_{DSS} = 2 to 20 mA
V_p = −0.5 to −7.5 V
g_m = 2 to 6.5 mS

In practice, the reverse biased gate-channel junction is not ideal; a leakage current I_{GSS} flows out of the gate terminal of an n-channel JFET (into the gate for a p-channel device). In common with other junction leakage currents, I_{GSS} is very temperature dependent and doubles approximately every 10 °C.

Operation of a p-channel JFET is exactly the same as for an n-channel device except that channel and gate doping are interchanged, channel conduction is by holes instead of electrons, and the direction of drain current and the polarity of applied voltages are reversed.

p-channel JFETs?

The symbol for a JFET must distinguish not only between n-channel and p-channel devices but also between JFETs and BJTs. Figure 7.3 shows the JFET symbols in common usage together with current and voltage directions in normal operation.

The JFET is called a **depletion** device since a conducting channel exists between source and drain under zero gate-source voltage. Application of V_{GS} with the correct polarity depletes or narrows the channel. This is distinct from the MOST which can be either depletion or **enhancement**, as discussed later.

Characteristic equations

JFET operation in the ohmic region can be characterized by the equation

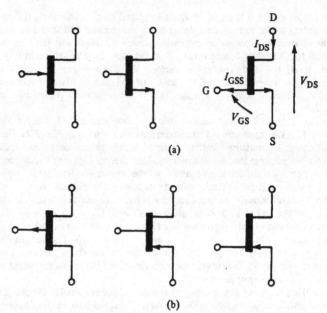

Fig. 7.3 Symbols for JFETs: (a) n-channel, (b) p-channel.

These equations are valid approximations to the large-signal d.c. behaviour for which device capacitance may be ignored.

$$I_{DS} = \frac{2I_{DSS}}{V_p^2}\left[V_{DS}\left(V_{GS} - V_p\right) - \frac{V_{DS}^2}{2}\right] \tag{7.1}$$

which contains an implicit dependence of I_{DS} on both V_{GS} and V_{DS}. Taking the ratio of I_{DS} to V_{DS} gives the effective d.c. channel conductance (G_{DS}) of the JFET channel (rather than its dynamic slope conductance)

$$G_{DS} = \frac{I_{DS}}{V_{DS}} = \frac{2I_{DSS}}{V_p^2}\left[V_{GS} - V_p - \frac{V_{DS}}{2}\right] \tag{7.2}$$

The d.c. channel resistance (R_{DS}) is the reciprocal of G_{DS}:

$$R_{DS} = \frac{V_{DS}}{I_{DS}} = \frac{V_p^2}{2I_{DSS}}\left[V_{GS} - V_p - \frac{V_{DS}}{2}\right]^{-1} \tag{7.3}$$

At $V_{DS} = 0\,\text{V}$,

$$R_{DS} = \frac{V_p^2}{2I_{DSS}}\left[V_{GS} - V_p\right]^{-1} \tag{7.4}$$

$$= R_{DSO}\left[1 - \frac{V_{GS}}{V_p}\right]^{-1} \tag{7.5}$$

where

$$R_{DSO} = R_{DS}\,|_{V_{GS}=0} = \frac{-V_p}{2I_{DSS}} \tag{7.6}$$

R_{DSO} is the minimum drain-source resistance (or ON-resistance) of the JFET. R_{DS} has particular relevance to voltage-variable resistor and switching applications of FETs which are covered later in this chapter.

In the saturation region a JFET is described by

$$I_{DS} = I_{DSS} \left[1 - \frac{V_{GS}}{V_p} \right]^2 \qquad (7.7)$$

In practice, the index can vary between 1.9 and 2.1.

which shows the ideal constancy of I_{DS} against V_{DS} in this region and also the square-law relationship between I_{DS} and V_{GS} which is of particular significance in communications circuits, reducing cross-modulation and intermodulation.

Differentiating Equation 7.7 with respect to V_{GS} results in the dynamic transconductance (g_m), given by

$$g_m = \frac{dI_{DS}}{dV_{GS}} = \frac{-2I_{DSS}}{V_p} \left[1 - \frac{V_{GS}}{V_p} \right] \qquad (7.8)$$

At $V_{GS} = 0$ V, the transconductance is at a maximum

$$g_{mo} = \frac{-2I_{DSS}}{V_p} \qquad (7.9)$$

and

$$g_m = g_{mo} \left[1 - \frac{V_{GS}}{V_p} \right] \qquad (7.10)$$

Note that, since I_{DSS} and V_p have opposite signs, g_m and g_{mo} are always positive.

This parameter is of particular importance in small-signal a.c. amplifier circuits.

(The locus of the boundary between the ohmic and saturation regions is

$$V_{DS} = V_{GS} - V_p \qquad (7.11)$$

This can be substituted into Equation 7.1 to give the device equation for the boundary and beyond (Equation 7.7) since, in the saturation region, I_{DS} is ideally independent of V_{DS}.)

The MOST

Structure and operation

An n-channel MOST (Fig. 7.4a) is fabricated by diffusing n^+ source and drain regions into a p-type substrate (or body). The surface of the substrate between source and drain is covered by a very thin (usually less than 1000 Å) insulating layer of silicon oxide on top of which is laid the gate metallization (usually of aluminium).

The MOST is a very simple device requiring only one diffusion and is self-isolating within the substrate; adjacent devices do not interact. Saving of chip area due to the absence of an isolation diffusion (necessary with bipolar devices) is further enhanced by the very small device dimensions which are possible (source-drain separation of less than 1 μm). Thus many thousands of MOSTs can be fabricated within a very small area to form a complex circuit such as a microprocessor or memory array.

No apology is made for the mixture of units of length as those quoted here are in everyday use in the semiconductor industry.

Figure 7.4a shows that, in equilibrium, there is no channel connecting source and drain. If $V_{DS} = 0$ V and V_{GS} is taken positive, holes in the substrate are repelled from the surface below the gate and when V_{GS} reaches a sufficiently positive threshold

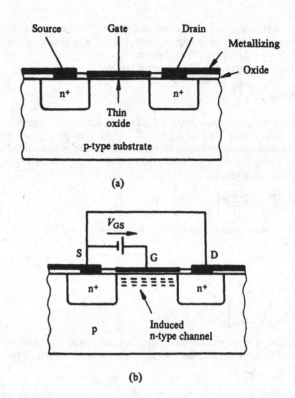

Fig. 7.4   n-channel enhancement MOST: (a) in equilibrium, (b) with $V_{GS} > V_{TH}$ showing induced channel.

V_{TH} was as high as 5 V for early devices but is lower than 0.9 V for modern MOSTs.

MOSTs can be either enhancement or depletion devices; JFETs are depletion only.

A different method of MOST construction, the *silicon gate* process (see Morant (1990), Chapter 3) has the advantages over the simple structure described here in that threshold voltages are reduced and alignment of the gate between source and drain is automatic.

voltage (V_{TH}), electrons are attracted predominantly from the source and drain regions to form an n-type channel between source and drain. This is called an **inversion** layer (Fig. 7.4b) where the normally p-type substrate has been inverted to n-type by the attraction of mobile electrons to the surface. Further increase of V_{GS} beyond V_{TH} merely increases the electron density at the surface thus increasing channel conductivity. This is called enhancement of the channel.

Unlike the JFET which is a depletion device, a MOST can be fabricated as either a depletion or an enhancement device depending on whether or not a channel is implanted (in processing) between source and drain. For an enhancement MOST (with no implanted channel), there is no source-drain conduction for $V_{GS} = 0$ V and I_{DSS} ideally is zero although, in practice, it is at leakage level. Increasing V_{GS} from zero first establishes the channel at $V_{GS} = V_{TH}$ and then, for $V_{GS} > V_{TH}$, enhances the channel. This is clearly illustrated in the transfer characteristic of Fig. 7.5a.

As with a JFET, for low values of V_{DS} a MOST operates in the ohmic region with saturation occurring for higher V_{DS} (see Fig. 7.5b). The only difference between the output characteristic of an enhancement MOST and that of a depletion JFET is the existence of channel conduction at $V_{GS} = 0$ V for the depletion device.

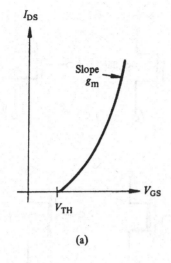

(a)

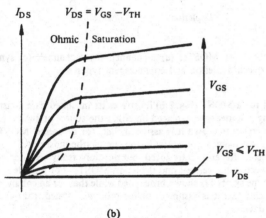

(b)

Fig. 7.5 n-channel enhancement MOST characteristics: (a) transfer characteristic in saturation region, (b) output characteristic.

Depletion MOSTs are fabricated by implanting a channel layer between the source and drain regions so that source-drain conduction occurs at $V_{GS} = 0$ V. Such a device is capable of both depletion and enhancement since there is no p–n junction to forward bias and thereby destroy the high input resistance. P-channel MOSTs are fabricated in a similar way to n-channel MOSTs except that the substrate is n-type and the source and drain regions are p^+. Circuits containing both n-channel and p-channel MOSTs have significance as CMOS (complementary MOS) which is considered later.

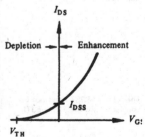

135

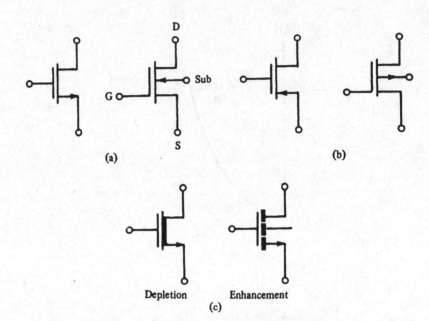

(a)

(b)

Depletion   Enhancement

(c)

Fig. 7.6 Symbols for MOSTs: (a) n-channel, (b) p-channel, (c) symbols which distinguish between depletion and enhancement types.

The symbol for a MOST (Fig. 7.6) illustrates its insulated gate construction; the arrow denoting channel type is placed on either the source or substrate lead. If no substrate connection is shown it is assumed that, for n-channel MOSTs, it is connected to the most negative potential in the circuit (the most positive potential for p-channel devices). Recently introduced, but not universally accepted, symbols discriminate between enhancement and depletion MOSTs. For a depletion device the channel bar in the symbol is shown broadened while that for an enhancement device is broken. In this text it is assumed, unless otherwise stated, that all MOSTs are enhancement devices.

Handling precautions.
A very small electrostatic charge is sufficient to produce a field that breaks down the gate insulation.

The insulated gate nature of the MOST provides a very high input resistance at the gate – of the order of $10^{18}\,\Omega$. Great care must be taken in handling MOS devices since electrostatic charge can easily destroy them.

Characteristic equations

In the ohmic region a MOST is characterized by the device equation

$$I_{DS} = \beta V_{DS}\left[(V_{GS} - V_{TH}) - \frac{V_{DS}}{2}\right] \tag{7.12}$$

The β for a MOST must not be confused with the β of a BJT!

where the device constant (β, in units of S/V) is given by

$$\beta = \mu\frac{\varepsilon_r\varepsilon_o}{t}\frac{w}{l} \tag{7.13}$$

136

Process parameters:

ε_r = relative permittivity of silicon oxide (≈ 4)
ε_o = permittivity of free space (8.85×10^{-12} F/m)
μ = majority carrier mobility (≈ 0.05 m^2/Vs for surface electrons)
t = oxide thickness

Design variables:
w = channel width
l = channel length

Since the boundary between the ohmic and saturation regions is described by

$$V_{DS} = V_{GS} - V_{TH} \tag{7.14}$$

the characteristic equation for the saturation region may be derived from Equation 7.12 as

$$I_{DS} = \frac{\beta}{2}\left[V_{GS} - V_{TH}\right]^2 \tag{7.15}$$

Equation 7.15 shows the square-law nature of the MOST transfer characteristic and, through β, the dependence of channel conductivity on device geometry; a short, wide channel has a higher conductance than a long, narrow channel with the same voltages applied to the device. This feature is used in a later section to design MOST logic gates.

As with JFETs the characteristic equations of MOSTs apply to both n-channel and p-channel devices provided that voltages and currents are given the correct algebraic signs.

Small-signal model

The behaviour of both JFETs and MOSTs for small a.c. signals can be represented by the equivalent circuit of Fig. 7.7. This equivalent circuit is primarily a voltage-controlled current source ($g_m v_{gs}$) with r_{ds} modelling the output slope resistance. The transconductance (g_m) can either be calculated from Equation 7.8 or obtained from manufacturers' data, as can the value of r_{ds}. At low frequencies, only these two components are necessary.

For high-frequency applications the gate-channel capacitance is represented by

For MOSTs, β is a measure of the source-drain conductivity.

Owing to surface effects, carrier mobility in the channel is usually less than half that in bulk material.

Comparing the JFET Equations 7.1 and 7.7 with the MOST Equations 7.12 and 7.15, it is seen that they are very similar in form.

$$g_m = \frac{dI_{DS}}{dV_{GS}} = \beta[V_{GS} - V_{TH}]$$

$$= \beta\sqrt{\frac{2I_{DS}}{\beta}} \quad \text{from (7.15)}$$

$$= \sqrt{2\beta I_{DS}}$$

Hence $g_m \propto \sqrt{I_{DS}}$ for an FET ($\propto I_{DS}$ for BJT).

For n-channel depletion FETs, I_{DS} and V_{DS} are positive, V_{GS} and V_p are negative. V_{TH} is positive for an n-channel enhancement device while V_{GS} may be either positive or negative. The opposite polarities apply for p-channel FETs.

See Sparkes (1987), Chapter 3.

On data sheets the capacitances C_{iss} and C_{rss} are often quoted.

$$C_{gs} = C_{iss} - C_{rss}$$

and $C_{gd} = C_{rss}$ if $C_{ds} \ll C_{gd}$. Also $r_{ds} = 1/g_{os}$.

In an actual device the gate-channel capacitance is distributed along the channel length.

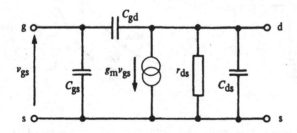

Fig. 7.7 FET small-signal a.c. equivalent circuit.

two (lumped) components, C_{gs} and C_{gd}; C_{ds} is the stray capacitance between drain and source.

Typical values of the equivalent circuit parameters for a JFET are $g_m = 5\,mS$, $r_{ds} = 20\,k\Omega$, $C_{gs} = C_{gd} = 4\,pF$, $C_{ds} = 1\,pF$.

FET amplifiers

FETs are particularly useful as low-noise devices in high input resistance amplifiers. Before we consider amplifier circuits, the biasing of a FET must be investigated. In this treatment, an n-channel (depletion) JFET is assumed but the techniques described are applicable equally to p-channel and to enhancement devices.

Biasing

Although shunt feedback biasing can be employed, the consequent reduction of input resistance destroys one of the major assets of FETs in amplifier circuits. To preserve a very high input resistance the following techniques may be used.

Voltage bias. In the basic common-source configuration of Fig. 7.8a the source is connected directly to earth and the gate-source voltage is established by connecting the gate via a high-valued resistor (R_G) to a negative supply voltage (V_{GG}). If the gate leakage current (I_{GSS}) is zero,

$$V_{GS} = V_{GG} \tag{7.16}$$

which is the equation of a bias line shown superimposed on the transfer characteristic limits in Fig. 7.8b. Owing to spread of the characteristic (V_p and

R_G is present to avoid short-circuiting the input signal to earth.

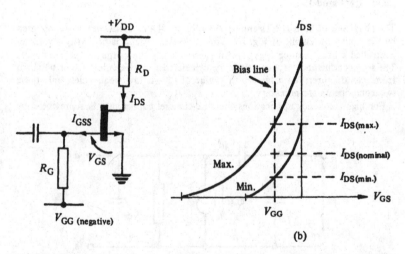

(b)

Fig. 7.8 Voltage bias circuit (a) with bias line superimposed on transfer characteristic (b).

138

I_{DSS}), the quiescent drain current (I_{DS}) is subject to wide variation from device to device, ranging from $I_{DS(min)}$ to $I_{DS(max)}$ as shown. At temperatures where the voltage dropped across R_G due to I_{GSS} is significant, the bias line equation is modified to

$$V_{GS} = V_{GG} + I_{GSS}R_G \tag{7.17}$$

which introduces a further variability of bias current.

Worked Example 7.1

An n-channel JFET is voltage biased (Fig. 7.8a) with $V_{GG} = -1$ V and $R_G = 1$ MΩ. If $I_{GSS} = 10$ nA at 25 °C, comment on its significance at both 25 °C and 125 °C.

Solution. At 25 °C, $I_{GSS} \times R_G = 10^{-9} \times 10^6 = 1$ mV, which certainly is insignificant compared with $V_{GG} = -1$ V. $V_{GS} = -999$ mV.

Remembering that I_{GSS} doubles approximately every 10 °C, there are ten doublings between 25 °C and 125 °C which is a factor of 2^{10} or approximately 10^3. Therefore, at 125 °C, I_{GSS} becomes $10^3 \times 10^{-9} = 1$ μA and $I_{GSS} \times R_G = 1$ V. V_{GS} is now approximately 0 V and $I_{DS} = I_{DSS}$, a significant departure from the design value ignoring I_{GSS}.

If V_{GS} is calculated as being positive (for an n-channel JFET) the only inference that can be made is that the gate is forward biased and the high input resistance is not obtained.

Automatic (or self) bias. As in BJT biasing, in order to aid bias stability, a resistor (R_S) can be introduced into the source lead (Fig. 7.9a). Again the gate is returned to a voltage supply (V_{GG}) via resistor R_G.

The bias line equation is now

$$I_{DS}R_S + V_{GS} = V_{GG} + I_{GSS}R_G$$

or $$I_{DS} = -\frac{V_{GS}}{R_S} + \frac{V_{GG}}{R_S} + I_{GSS}\frac{R_G}{R_S} \tag{7.18}$$

Here a graphical approach is used. Alternatively, the device Equation 7.7, with maximum and minimum I_{DS} plus I_{DSS} and V_p limits, can be manipulated to yield V_{GG} and R_S.

and is shown plotted against the transfer characteristic in Fig. 7.9b; its slope is $-1/R_S$. Several observations can be made. First, considering the case where $V_{GG} = 0$ V and assuming $I_{GSS} = 0$, it is evident that, due to the finite slope of the bias line compared with that in the voltage bias circuit, the difference between the maximum and minimum limits of quiescent I_{DS} is reduced. Second, for the same design value of drain current, if V_{GG} is taken positive and R_S correspondingly increased, the spread of bias current around its nominal value becomes smaller at the expense of a higher voltage drop across R_S. The positive gate supply voltage can be derived by a resistive potential divider between the drain supply (V_{DD}) and earth.

There is a trade off between bias instability and the fraction of the supply utilized for bias purposes.

If I_{GSS} is significant, the drain bias current is affected as shown by Equation 7.18 but for a given value of R_G (which, in the case of a potential divider, is the parallel combination of the two resistors) the effect is reduced with increasing R_S.

This automatic bias circuit is preferred for biasing FETs whether n-channel or p-channel, depletion or enhancement. For common-source operation, R_S must be capacitively decoupled otherwise the series negative feedback reduces the signal voltage gain.

Design Example 7.1

Figure 7.10 shows the transconductance spread for an n-channel JFET. Bias the transistor in a common-source configuration with a drain current of between 3 and 5 mA. The input resistance (R_G) is to be 1 MΩ.

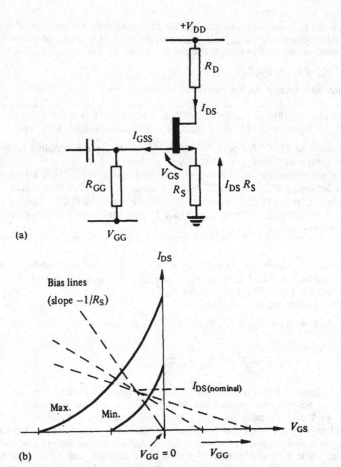

Fig. 7.9 Automatic bias circuit (a) with several bias lines (increasing V_{GG} and R_S) (b).

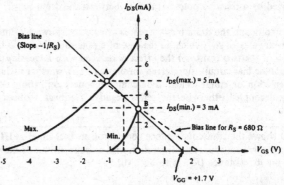

Fig. 7.10 n-channel JFET transfer characteristic spread.

Solution. If a voltage bias circuit (Fig. 7.8) is considered, V_{GG} would have to be 0 V to satisfy the $I_{DS(min)}$ specification but, in that case, $I_{DS(max)}$ would be 8 mA. Therefore, the automatic bias circuit of Fig. 7.9 should be used.

First, decide on which bias circuit to use ...

The bias line must be drawn joining the intersections (points A and B) of the I_{DS} specification limits with the extremes of the transfer characteristic spread, projecting it to cut the V_{GS} axis. From the graphical construction

... then draw the bias line.

$$V_{GG} = +1.7 \text{ V}$$

$$R_S = 567 \,\Omega$$

Remember that the slope of the bias line is $-1/R_S$.

and, for $I_{DS} = 4$ mA, $V_{GS} = -0.5$ V (estimated from the characteristic). This is the minimum value of R_S which meets the specification; a higher preferred value must be used and $R_S = 680 \,\Omega$ is selected. This value also accommodates resistor tolerances.

In assigning resistor values, attention must be paid to preferred values and tolerances. When a preferred value is chosen for a component, initial calculations usually have to be modified.

V_{GG} must now be recalculated. The new bias line drawn through $V_{GS} = -0.5$ V, $I_{DS} = 4$ mA with a slope of $-1/680$ intersects the V_{GS} axis at $V_{GG} = +2.2$ V. (This does not need to be drawn on the characteristic; coordinate geometry yields the answer.)

Therefore the circuit of Fig. 7.9a with $V_{GG} = +2.2$ V, $R_S = 680 \,\Omega$ and $R_G = 1\text{M}\Omega$ meets the specification.

If a single, V_{DD}, supply is to be used with two resistors forming a potential divider to generate V_{GG}, their values must be calculated from

$$V_{GG} = V_{DD}\frac{R_2}{R_1 + R_2} = +2.2 \text{ V}$$

and

$$R_G = R_1 \| R_2 = 1 \text{ M}\Omega$$

In both cases, to realize a common-source configuration for a.c. signals, R_S must be decoupled with a capacitor.

Current sources

Like a BJT, a FET biased in common-source behaves as a current source with a moderately high output resistance (r_{ds}). If gate and source are connected together, the drain current is I_{DSS} provided that $|V_{DS}|$ is greater than $|V_p|$ or $|V_{TH}|$. A much higher output resistance is achieved when an undecoupled source resistor R_S is included; in this case I_{DS} is less than I_{DSS}. Since in a d.c. current source circuit no a.c. signal is applied to the gate, the gate resistor (R_G) may be made a short-circuit.

Prove that the a.c. output resistance of a FET biased with an undecoupled source resistor R_S is

Exercise 7.1

$$r_{out} = R_S + (1 + g_m R_S)r_{ds} \tag{7.19}$$

Temperature stability

JFET transfer characteristics are temperature dependent but exhibit the curious phenomenon of a certain (I_{DS}, V_{GS}) point being independent of temperature.

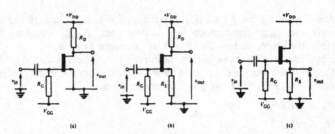

Fig. 7.11 FET amplifiers: (a) common-source, (b) series feedback, (c) source-follower.

Unfortunately this zero-temperature-coefficient bias point occurs rather close to pinch-off where I_{DS} and g_m are usually too small to be of value.

Common-source amplifier

The circuit of a common-source FET amplifier is shown in Fig. 7.11a. It is left to the reader to construct the equivalent circuit using the model of Fig. 7.7 (with the capacitors omitted) and to prove that

$$A_V = \frac{v_{out}}{v_{in}} = -g_m \frac{r_{ds} R_D}{r_{ds} + R_D} \tag{7.20}$$

$$\approx -g_m R_D \quad \text{if } r_{ds} \gg R_D \tag{7.21}$$

Also, $r_{in} = R_G$.

Series feedback amplifier

If series negative feedback is introduced by including an undecoupled source resistor (R_S) as shown in Fig. 7.11b, the voltage gain is reduced to

$$A_V = -\frac{g_m R_D}{1 + g_m R_S} \quad \text{(ignoring } r_{ds}) \tag{7.22}$$

$$\approx -\frac{R_D}{R_S} \quad \text{if } g_m R_S \gg 1 \tag{7.23}$$

And, again, $r_{in} = R_G$.

   In design, the gate resistor (R_G) should be chosen to meet the input resistance specification, provided that the effect of I_{GSS} and its temperature dependence does not unduly upset the bias conditions. In order to meet stringent bias stability and high input resistance specifications it may be necessary to use the bootstrap bias technique (see Chapter 4) which enhances a low-valued bias resistance (for d.c. stability) into a high a.c. signal input resistance.

Exercise 7.2   In a FET series feedback amplifier, $R_D = 4.7\,k\Omega$ and $R_S = 1\,k\Omega$. If the g_m of the device is $2\,mS$ and $r_{ds} = 20\,k\Omega$, determine the small-signal voltage gain of the circuit.
[*Answer*: $A_V = -3.13$. Here, $g_m R_S$ is only 2.]

If R_S is capacitively decoupled, calculate the voltage gain of the resultant common-source amplifier.

[*Answer*: $A_V = -7.61$]

Source follower

Exercise 7.3

If the output signal is taken from the source terminal, the source-follower circuit of Fig. 7.11c results. The voltage gain of this circuit is given by

$$A_V = \frac{g_m R_S}{1 + g_m R_S} \to 1 \quad \text{if } g_m R_S \gg 1 \tag{7.24}$$

Since the g_m for a FET is usually much lower than that of a BJT, any attempt to use a high value of R_S would appear to prove useful in achieving a voltage gain close to unity. However, for a constant voltage across R_S, as R_S is increased so the drain bias current must decrease and g_m also decreases in sympathy. A maximum $g_m R_S$ product usually is realized for only moderate values of R_S, and the voltage gain is unlikely to exceed 0.9. The solution to this problem is to replace R_S with a high output resistance current source thus achieving a high effective resistance in the source with a minimal d.c. voltage drop.

There is a striking similarity between the voltage gain equations for FET and BJT circuits. It is not necessary to remember both sets of equations provided they are in g_m form ($g_m = 1/r_e$). FETs do not have a parameter corresponding to r_e, but the g_ms are equivalent in every respect.

It can be proved that the output resistance of the source-follower is:

$$r_{out} = R_S \| r_{ds} \| \frac{1}{g_m} \approx \frac{1}{g_m} \tag{7.25}$$

which, again owing to the relatively low g_m of a FET, is significantly higher than the output resistance of the emitter-follower counterpart.

The input resistance of the simple source-follower is equal to R_G but bootstrap bias techniques can raise this value.

Prove the voltage gain expressions for the common-source, series feedback and source-follower amplifiers given in Equations 7.20, 7.22 and 7.24. Also prove that the output resistance of a source-follower is as given in Equation 7.25.

Exercise 7.4

Differential amplifier

Matched FETs can be used in a differential amplifier configuration to achieve a very high a.c. input resistance ($10^{12}\,\Omega$) and extremely low d.c. input currents (30 pA).

FET and bipolar technologies have been combined in BIFET operational amplifiers, the FET input stage improving the input performance compared with all-bipolar circuits. Operational amplifiers using only MOSTs are also available.

Although MOSTs inherently exhibit much higher input resistances, this figure of 10^{12} ohms is due almost entirely to reverse biased gate diodes incorporated in the MOSTs for protection purposes.

Voltage-variable resistor

The channel of a FET, whether JFET or MOST, is a region of semiconductor material whose depth, and hence resistance, can be controlled by the gate-source voltage. This is the essence of a voltage-variable resistor (VVR) or voltage-controlled resistor (VCR) which, for example, can be placed in either the series or shunt arm (or both) of a potential divider to perform a voltage-controlled attenuator function. Alternatively, the VVR can be connected in the emitter circuit of a series feedback amplifier to control the a.c. voltage gain.

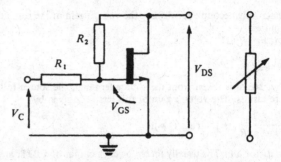

Fig. 7.12   FET VVR with linearizing feedback.

FET channel conduction is bidirectional but not necessarily symmetrical about the origin of the output characteristic. In the ohmic region the output characteristic exhibits significant curvature and it is evident that, for approximately linear behaviour and low distortion, the signal voltage amplitude (V_{DS}) across the channel must be restricted to several tens of millivolts.

Equation 7.3 clearly illustrates the dependence of d.c. channel resistance (R_{DS}) on both V_{GS} and V_{DS} but, if the control voltage can be modified to include an appropriate V_{DS}-dependent term, R_{DS} would then be linearized.

Consider the circuit of Fig. 7.12 in which a resistor R_1 is connected between the control voltage source (V_C) and the FET gate; also R_2 is connected between drain and gate providing some feedback of V_{DS} to the gate. For this circuit

$$\frac{V_{DS} - V_{GS}}{R_2} = \frac{V_{GS} - V_C}{R_1}$$

Therefore

$$V_{GS} = \frac{R_2 V_C + R_1 V_{DS}}{R_1 + R_2}$$

If $R_1 = R_2$ then

$$V_{GS} = \frac{V_C}{2} + \frac{V_{DS}}{2}$$

which, when substituted into Equation 7.3, gives

$$R_{DS} = \frac{V_p^2}{2I_{DSS}} \left[\frac{V_C}{2} + \frac{V_{DS}}{2} - V_p - \frac{V_{DS}}{2} \right]^{-1}$$

$$= R_{DSO} \left[1 - \frac{V_C}{2V_p} \right]^{-1} \tag{7.26}$$

which is independent of V_{DS}! The minimum drain-source resistance (R_{DSO}), at $V_C = 0$, is given by Equation 7.6 as

$$R_{DSO} = -\frac{V_p}{2I_{DSS}}$$

Several constraints imposed by this circuit are recognized:

1. The control sensitivity has been halved due to the feedback.
2. To achieve maximum variation of resistance, R_1 and R_2 should be high in value since the total resistance across the source and drain terminals of the VVR is R_{DS} in parallel with $(R_1 + R_2)$.
3. Also, the source terminal of the FET must be at a fixed potential (for example, earth potential) unless the control voltage source can float with the signal.

FET switches

Due to the ease with which FETs can be turned OFF and ON they have wide application both as shunt and series switches. A typical example of the application of a discrete FET shunt switch is to discharge a capacitor; a multiplicity of such switches are used in digital-to-analogue and analogue-to-digital converters. A series switch normally is used as a transmission gate to conditionally connect a signal from one part of a circuit or system to another. In some devices an ON resistance as low as 1 Ω can be achieved but, since such devices are physically large with a correspondingly high capacitance, their switching speed is rather limited. Smaller devices with ON resistances in the range 50 to 200 Ω can be switched much faster (tens of nanoseconds or less).

Bannister and Whitehead (1991, Chapter 4) illustrate several applications of FET switches.

Shunt switch

First, considering an n-channel depletion FET, the ON state ($R_{DS} = R_{DSO}$) is readily achieved by making $V_{GS} = 0$ V. To turn OFF the switch, V_{GS} must be taken more negative than the most negative signal excursion of source or drain by at least the magnitude of the pinch-off voltage (V_p) or threshold voltage (V_{TH}).

Now, for an n-channel enhancement MOST the switch is OFF if $V_{GS} = 0$ V. To turn ON the switch, V_{GS} must be driven positive by an amount (certainly exceeding V_{TH}) limited by the gate breakdown voltage.

For p-channel devices, the polarity of the gate voltage simply is reversed.

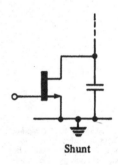

Shunt

Since a FET is bidirectional, source and drain terminals reverse rôle if their normal polarities are reversed.

Series switch

Controlling the ON and OFF states of an enhancement device used as a series switch is straightforward. Provided that the gate breakdown voltage is not exceeded in either direction, the ON state is achieved (for an n-channel MOST) by driving the gate to a high positive potential; for the OFF state, the gate is taken more negative than the most negative signal excursion.

Controlling an n-channel depletion FET in a series switch configuration is rather more complex: the gate control voltage must float with the signal to achieve a constant ON resistance and, in the case of an n-channel JFET, the gate must not be forward biased. A simple circuit which automatically provides the correct turn-ON gate voltage is shown in Fig. 7.13.

If the control voltage (V_C) is driven above the input voltage (V_{in}), the diode is reverse biased and the resistor (R), connected between gate and source, establishes $V_{GS} = 0$ V – the ON state. If V_C is taken negative (at least $|V_p|$ below V_{in}), the diode conducts pulling the gate negative with respect to the source and turning OFF the switch.

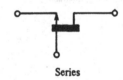

Series

It is V_{GS} which controls the state of the switch.

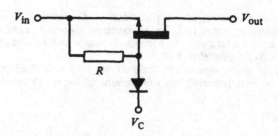

Fig. 7.13　FET series with drive isolation.

A suitable practical value for R is 10 kΩ.

The value of R is not critical but, if too high, the turn-ON time of the switch is long and, if too low, a significant load may be placed on the input and control circuitry.

FET switches have the advantage over their BJT counterparts in that they have no equivalent of the offset voltage $V_{CE(sat)}$. The FET channel is resistive; as there are no junctions in the conducting path, the voltage developed across an ON FET switch is the product of R_{DSO} and the current flowing.

MOS logic gates

It is customary to use depletion MOSTs as active loads (see Morant (1990), Chapter 4) but design of such gates is rather more complex than for the all-enhancement MOST gate design. The importance of the example given here lies in simple illustration of the design process and the resulting transistor geometries.

In Chapter 5 the concept of using BJTs as active loads in amplifier circuits was introduced. The attraction of this approach is due to area constraints imposed on an integrated circuit chip by which it is economic to replace, wherever possible, resistors with small area transistors. MOS circuitry is no exception; logic gates can be constructed using MOSTs and no other components. In the following circuits, n-channel enhancement devices exclusively are considered although the concepts apply equally for p-channel devices.

MOS inverter

The simplest logic gate is an inverter, shown in Fig. 7.14a and comprising two MOSTs – a driver transistor and a load transistor. The load MOST has its gate and drain terminals connected together ($V_{DS} = V_{GS}$). Referring to Fig. 7.14b, it is evident that the $I - V$ characteristic of this device corresponds to that of a nonlinear resistor and lies within the saturation region of operation.

We shall now design an inverter gate against a given specification which permits calculation of the physical geometries of the two transistors.

Specification. OFF state (both transistors OFF): $V_{in} = +0.5$ V, $V_{out} = +4.0$ V.

ON state (both transistors conduct): $V_{in} = +4.0$ V, $V_{out} = +0.5$ V, $I_{DS(ON)} = 50 \mu$A.

Assume $V_{TH} = +1.0$ V for both transistors.

This is an assumption used to simplify the design. In the actual circuit, the common substrate is connected to earth, as is the source of the driver transistor. Since the source of the load transistor is not at this potential, the V_{TH}s of the transistors differ.

Design. (a) Considering the OFF state:

$V_{in(OFF)} = +0.5$ V

Driver $I_{DS} \approx 0$

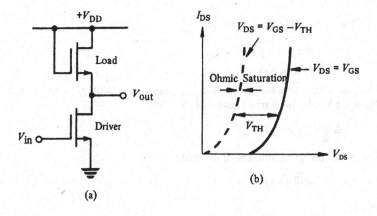

(a)

(b)

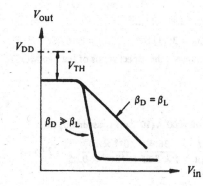

Fig. 7.14 (a) MOS inverter. (b) Load MOS characteristic. (c) Gate transfer characteristic.

Only leakage current flows but is sufficient to make $V_{GS} \approx V_{TH}$ for the load transistor. Therefore

$$V_{DD} = V_{out(OFF)} + V_{TH(load)}$$
$$= 4.0 + 1.0\,V$$
$$= +5\,V$$

(b) In the ON state, the driver transistor is in its ohmic region ($V_{DS} = +0.5\,V$ and $V_{GS} = +4.0\,V$) and Equation 7.12 applies:

$$I_{DS} = \beta_D \left[V_{DS}(V_{DS} - V_{TH}) - \frac{V_{DS}^2}{2} \right]$$

where β_D is the device constant of the driver MOST.

147

$$5 \times 10^{-5} = \beta_D \left[0.5(4-1) - \frac{0.5^2}{2} \right]$$

$$= 1.375 \times \beta_D$$

$$\beta_D = 36.4 \, \mu S/V$$

The load MOST is in its saturation region (always) and, with $I_{DS} = 50 \, \mu A$, $V_{GS} = V_{DS} = 4.5 \, V$ substituted into Equation 7.15:

$$I_{DS} = \frac{\beta_L}{2} \left(V_{GS} - V_{TH} \right)^2$$

where β_L is the device constant of the load MOST,

$$5 \times 10^{-5} = \frac{\beta_L}{2} (3.5)^2$$

Therefore

$$\beta_L = 8.16 \, \mu S/V$$

The above calculations have yielded:

$$V_{DD} = +5 \, V \qquad \beta_D = 36.4 \, \mu S/V \qquad \beta_L = 8.16 \, \mu S/V$$

For given process parameters, the aspect ratios of the two MOSTs can be determined from Equation 7.13.

$$\beta = \mu \frac{\epsilon_r \epsilon_o}{t} \frac{w}{l}$$

If an oxide thickness of 1000 Å $(10^{-7} \, m)$ is assumed, then

The designer simply has to determine the length and width of the transistors consistent with the constraints of the process.

$$\frac{w}{l} \, (\text{driver}) = \frac{\beta_D t}{\epsilon_r \epsilon_o \mu} = \frac{36.4 \times 10^{-6} \times 10^{-7}}{4 \times 8.85 \times 10^{-12} \times 0.05} = 2.056$$

and

$$\frac{w}{l} \, (\text{load}) = \frac{\beta_L t}{\epsilon_r \epsilon_o \mu} = 0.461$$

The driver MOST (the lower resistance device) is relatively short and wide compared with the higher resistance load MOST which is long and thin. It remains only to determine the actual channel dimensions within the constraints of device area and process limitations.

The β ratio reduces to

$$\frac{\beta_D}{\beta_L} = \frac{w_D}{l_D} \frac{l_L}{w_L}$$

Millman and Grabel (1987, pp. 229–230) discuss these two modifications.

The ratio β_D / β_L influences the logic swing ($V_{out(OFF)} = V_{out(ON)}$) of the circuit and, in order to achieve a high level of discrimination between the two output states, β ratios as high as 20:1 are sometimes used. It is possible to operate the enhancement load MOST in its ohmic region and thus achieve a $V_{out(OFF)}$ level very close to V_{DD} rather than being a threshold voltage lower. However, for this mode of operation, it is necessary to provide an extra voltage supply ($V_{GG} > V_{DD}$) to which the gate of the load MOST is connected. The expense of this extra supply is significant and most systems favour single-supply circuits.

If a depletion load MOST is used with an enhancement driver MOST, the logic swing is approximately from earth to V_{DD}. In this case the fabrication process must be capable of realizing both enhancement and depletion devices on a chip.

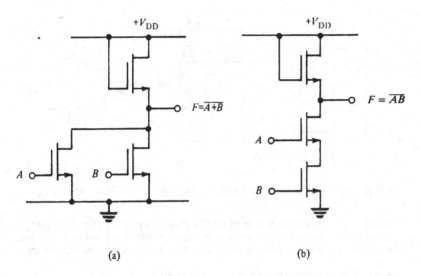

Fig. 7.15   (a) MOS NOR gate. (b) MOS NAND gate.

MOS NOR and NAND gates

Figure 7.15 shows the circuits of simple, two-input NOR and NAND gates using n-channel enhancement MOSTs.

In the NOR gate circuit of Fig. 7.15a, the output F is at logic 0 (≈ 0 V) if either input A or B or both is at logic 1 ($\approx V_{DD}$). When both A and B inputs are at logic 0, F is at logic 1. This is the NOR function, $F = \overline{A + B}$. Design of this NOR gate is identical to that of the inverter described in the previous section. In the ON state, the βs of the driver MOSTs individually must be large enough in relation to the β of the load MOST to realize the logic 0 output voltage specification.

For the two-input NAND gate of Fig. 7.15b, the ON state is defined when both inputs A and B are at logic 1 ($F = \overline{AB}$). Since the two driver transistors are in series, in order to achieve the specified logic 0 output voltage, the β of each driver transistor must be twice β_D for the simple inverter. (With three inputs, the multiplying factor would be three, and so on.) In the NAND circuit the driver MOSTs are larger in area than for the NOR circuit and the latter configuration normally is preferred.

CMOS logic gates

A major limitation of n-channel (or p-channel) logic gates is that when the circuit is ON it passes supply current and dissipates power. In complementary MOS (or CMOS) circuits, which use both n-channel and p-channel devices, only one transistor is ON at a time and no steady current is drawn from the supply. Pulses of supply current do occur when the gates change state.

Again, depletion load devices can be used or the load MOST operated in its ohmic region.

Remember that MOST conductance is proportional to β.

Total power dissipation, as well as area, restricts chip complexity. Power density is the likely limiting factor as devices become smaller and smaller.

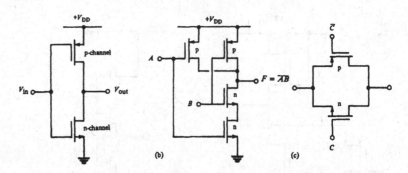

Fig. 7.16   (a) CMOS inverter. (b) CMOS NAND gate. (c) CMOS transmission gate.

Another very important feature of CMOS circuits is that they do not require a highly stable voltage supply; an unregulated supply of between 3 and 15 V (plus the low power dissipation) is particularly attractive for portable, battery-powered systems.

Figure 7.16 illustrates three simple CMOS circuits.

Logic 0 and 1 levels are very close, to earth and V_{DD} respectively.

In the CMOS inverter (Fig. 7.16a), when V_{in} is at logic 0 (0 V), the n-channel transistor is OFF while the p-channel transistor is ON providing an active **pull-up** for following circuitry; V_{out} is at logic 1 (V_{DD}). When V_{in} is at logic 1, the n-channel transistor is ON and the p-channel transistor is OFF; V_{out} is then pulled down to logic 0. The input voltage controls the state of both the driver and the load.

The two-input CMOS NAND gate of Fig. 7.16b extends this approach. Both inputs A and B must be at logic 1 to turn ON both series-connected n-channel devices and pull the output voltage down to logic 0; in this case both p-channel devices are OFF. Any other combination of inputs turns ON at least one p-channel transistor but the series connection of n-channel transistors is nonconducting and the output voltage is pulled to logic 1.

Figure 7.16c illustrates a very useful CMOS circuit, the analogue switch or transmission gate. Complementary (i.e. opposite polarity) gate control voltages (C and $\bar{C}$) are applied to the parallel n- and p-channel MOSTs. When C is at logic 1 (and $\bar{C}$ at logic 0), both transistors conduct with a low ON-resistance. Both transistors are OFF and the switch is open when C is at logic 0.

We have considered only simple MOS and CMOS circuits; much more complex circuits are in common use.

Summary

This chapter has attempted to present a simple introduction to the structure, operation and characterization of field-effect transistors, both JFETs and MOSTs, plus some of their many applications in linear and digital circuits.

We have seen how the drain current in a FET is a function not only of its gate-source voltage but also of the drain-source voltage. Ohmic and saturation regions of operation have been identified and expressed in terms of simple device equations which differ only in detail for depletion or enhancement devices, n-channel or p-channel. These equations provide a quantitative basis for the design of voltage-variable resistor circuits and of MOS logic gates.

Due to the wide spread of FET parameters such as V_p and I_{DSS}, d.c. biasing is not a simple matter. The concept of a bias line has been introduced to permit graphical design of two types of bias circuit. A small-signal FET equivalent circuit was used to analyse single-stage amplifiers which, although they find wide application for their high input resistance, suffer from the characteristically low transconductance of field-effect devices.

Being resistive rather than junction devices, FETs do not exhibit the equivalent of the $V_{CE(sat)}$ offset found in the BJT and are ideally suited to use as switches. The gate drive voltage requirements for ON and OFF states have been investigated for the various types of FET.

The most widespread use of FETs undoubtedly is in MOS and CMOS integrated circuits. Owing to the very small device dimensions, low power dissipation and the ability to replace resistors by transistors, extremely complex circuits containing hundreds of thousands of devices can be fabricated on a single silicon chip. We have investigated the design of simple MOS logic gates and have seen, in an example, how the device equations can be applied to yield the actual dimensions of the transistors.

Field-effect transistors are perhaps the most important components in modern electronics. Further study of this area will be amply rewarded.

Problems

7.1 On the transfer characteristic of a JFET, at what point does the tangent to the characteristic at $V_{GS} = 0$ V intersect the V_{GS} axis? What practical use can be made of this result?

7.2 An n-channel MOS process uses an oxide thickness of 500 Å. Given that the surface mobility of electrons is 0.05 m^2/Vs and that the relative permittivity of silicon oxide is 4, determine the channel width of a MOST whose device constant and channel length are 20 μS/V and 10 μm respectively.

7.3 An n-channel JFET has the transfer characteristic spread illustrated in Fig. 7.10 and is voltage biased with $V_{GG} = -0.5$ V and $R_G = 1$ MΩ.
(a) Assuming that I_{GSS} is zero, estimate the possible range of quiescent I_{DS}.
(b) If I_{GSS} is 250 nA at an elevated temperature, determine V_{GS} and the quiescent I_{DS} range.

7.4 Using the n-channel JFET transfer characteristic spread of Fig. 7.10, design an automatic bias circuit to yield a quiescent I_{DS} range of 3 to 6 mA. The effect of I_{GSS} may be ignored.

7.5 If $g_m = 4.5$ mS and $r_{ds} = 20$ kΩ, calculate the output resistance of the current source designed in Problem 7.4.

7.6 Calculate the small-signal voltage gain and output resistance of a source-follower with $R_S = 3.3$ kΩ. ($g_m = 3$ mS and $r_{ds} = 20$ kΩ.)

7.7 In a JFET differential amplifier the circuit of Problems 7.4 and 7.5 is used as the tail current source. If the g_m of the differential transistors is 3.5 mS and the drain load resistance is 2.2 kΩ, determine the differential voltage gain, common-mode gain and common-mode rejection ratio of the amplifier for single-ended output. (Ignore the r_{ds} of the differential transistors.)

7.8 For the MOS inverter of Fig. 7.14 prove that, provided both load and driver transistors operate in the saturation region,

$$V_{out} = -\sqrt{\left(\frac{\beta_D}{\beta_L}\right)}\,(V_{in} - V_{TH}) + V_{DD} - V_{TH}$$

where β_L and β_D are the device constants of the load and driver transistors respectively.

7.9 For a two-input MOS NOR gate using n-channel enhancement MOSTs, calculate the supply voltage and the device constants for the driver and load transistors required to meet the specification:

logic 1 level = +6 V
logic 0 level = +1 V
$I_{DS(ON)} = 100\,\mu A$

It may be assumed that all transistors have a threshold voltage of +2 V.

7.10 Design a three-input MOS NAND gate to the specification of Problem 7.9.

Audio power amplifiers

Objectives

□ To translate the performance requirements of an audio power amplifier into terms of power, voltage and current handling, gain precision and distortion.
□ To explain the function of complementary emitter-followers in the output stage.
□ To explain how the amplified diode is used to reduce crossover distortion in class AB, push-pull operation.
□ To investigate collector load bootstrapping.
□ To design a simple audio power amplifier with overall negative feedback.

An audio system

This chapter will relate particularly to a domestic high-fidelity (hi-fi) audio system although the reader must not forget professional applications such as public address systems (indoor and outdoor) as well as broadcast and recording studio monitors. In the hi-fi system shown in Fig. 8.1 the audio source is selected from a radio tuner (AM or FM), tape, vinyl disc, compact disc or microphone. It is played through two loudspeakers, one each for left and right channels in a stereophonic (stereo) system, or one in the case of a monaural (mono) system. The selected audio signal is processed in an amplifier which is invariably split into two sections: a preamplifier followed by a power amplifier, with two channels for stereo.

The preamplifier selects the input source from a variety of voltage levels and output impedances, e.g. a microphone providing 0.5 mV at 600 Ω or a compact disc unit 2 V at 100 Ω. Also there may be a requirement to provide equalization to compensate for pre-emphasis essential to the process of recording on vinyl disc. The first stage therefore bring inputs up to equal levels with a flat frequency response over the audio range (nominally 20 Hz to 20 kHz). Further amplification with tone controls to raise or lower the bass and treble ends of the response may appear in the form of a graphic equalizer or simple rotary or slide controls. High-frequency (scratch) and low-frequency (rumble) switched filters also may be provided. The output amplitudes of the left and right channels are either individually controlled or the gain controls are ganged together and a separate balance control provided. Thus the preamplifier provides to the power amplifier a variable signal level at low impedance. The power amplifier has no controls other than its power supply on/off switch and functions to amplify the preamplifier signal to drive the loudspeakers. Naturally, there is a power amplifier for each channel, accurately matched in terms of gain and frequency response (flat over the audio band).

Loudspeakers are manufactured in a wide range of power ratings from a few watts to hundreds of watts. Also they are available in a variety of impedances of which 4, 8 and 16 Ω are the most common. It is immediately obvious from the loudspeaker power and impedance specifications that non-trivial voltages and current are involved. Therefore the power amplifier must provide both voltage and current gain

At the recording stage, the low frequencies are attenuated (to avoid grooves touching) and must be accentuated at playback.

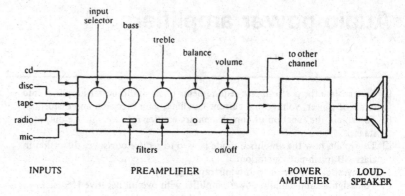

Fig. 8.1 Domestic hi-fi system.

over the audio spectrum with flat frequency response, and the gain should be accurately defined to permit close matching of two power amplifiers in a stereo system.

Voltage, current and power

In linear networks, sinusoidal waveforms remain sinusoidal with the same frequency and varying only in amplitude and phase.

This is Fourier series representation and is covered in O'Reilly (1989), Chapter 2.

At this stage it would be useful to review the 'properties' of sinusoidal electrical signals. Sinewaves are widely used in circuit analysis since they are easy to handle. Also, it should be remembered that any repetitive signal can be represented by a summation of harmonically related sinewaves. Consider a sinusoidal voltage v applied to resistance R (Fig. 8.2). By Ohm's law, a current i flows which is also sinusoidal and is of magnitude v/R.

v may be expressed as $v = V \sin \omega t$ and i as $i = I \sin \omega t$ where ω (in radians per second) is the angular frequency of the signal. ω is related to the frequency f (in hertz) by $\omega = 2\pi f$ and to the period T (in seconds) of the sinewave by $\omega = 2\pi T$. The peak amplitude of the waveform is V, that of the current is I, and the root-mean-square (rms) amplitudes are $V/\sqrt{2}$ and $I/\sqrt{2}$. Hence the power dissipated in R is

$$P = (V/\sqrt{2})(I/\sqrt{2}) = VI/2 = V^2/2R \tag{8.1}$$

Equation 8.1 relates the power dissipated in a resistive load, averaged over each cycle, to the peak voltage drive (V). Now the peak-to-peak voltage ($2V$) is a rather more

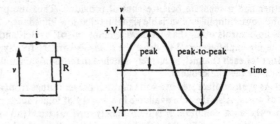

Fig. 8.2 Sinusoidal voltage applied to resistor.

154

useful consideration: it is the total voltage swing which must be provided by the power amplifier and, at maximum, is the total power supply voltage (V_{CC}) required for the amplifier.

This is not true if the amplifier output is transformer coupled to the load.

With $V_{CC} = 2V$ or $V = V_{CC}/2$, $P = V_{CC}^2/8R$ (8.2)

Usually, however, the output of an amplifier cannot be driven through the whole supply voltage; several volts are required for base-emitter voltage drops.

Here we have implied that the loudspeaker impedance is purely resistive. This is far from the truth since the wound nature of the voice coil immediately introduces inductance and inter-turn capacitance as well as resistance – a truly complex impedance further complicated by mechanical resonances. Therefore, since the voltage across a loudspeaker and the current through it are not in phase, it is not correct to call the power 'rms power'. It should be denoted 'apparent power' but is conventionally referred to as 'average power'.

Let us get an idea of the magnitudes of voltage and current with which we will have to deal.

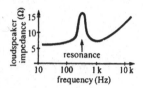

See Fidler and Ibbotson, Chapter 8.

For a 4 Ω loudspeaker (assumed resistive) operated at 50 W average power dissipation, determine the peak amplitudes of the sinusoidal voltage and current.

Solution. Since $P = V^2/2R$ (from Equation 8.1)

$V^2 = 2 \times 4 \times 50$

$\quad = 400$

giving

$V = 20$ V

This is the peak sinusoidal amplitude. Remember that the power amplifier total output voltage swing must be the peak-to-peak amplitude or 40 V and its power supply necessarily will be several volts greater than this.
The peak current $I = V/R = 5$ A.

Worked Example 8.1

The power is dissipated as heat and acoustic power.

If the 4 Ω loudspeaker dissipates 10 W, show that the peak-to-peak voltage is 18 V and that the peak current is 2.25 A.

Exercise 8.1

Distortion

It is very convenient to consider amplifiers, including power amplifiers, as being linear systems, i.e. the output signal is exactly proportional to the input signal – the utmost in high fidelity! For various reasons this cannot be achieved in practice. For example, the slope resistance r_e of a BJT depends on the **instantaneous** collector current (see Chapter 3), and therefore the voltage gain of a common-emitter amplifier is heavily dependent on the **instantaneous** signal amplitude: an input sinewave will not give a perfect sinewave at the output. This 'distortion' is reduced, but not eliminated completely, by negative feedback. Other sources of nonlinear distortion in amplifiers include voltage dependence of device parameters (e.g. the Early effect), crossover distortion (see later in this chapter) and poor regulation of the power supply

Refer to Chapter 9 of this text and Sparkes (1987), p. 127.

155

(see Chapter 9). Also, when very large signals are applied, transistors can saturate and cut off, and the output will not be able to follow the input.

It is not within the scope of this text to discuss distortion on a quantitative basis. However, it should be noted that the ear is very sensitive to distortion of a signal. Distortion is measured by filtering from the distorted (approximate) sinewave the fundamental sinewave component; the residual rms voltage, expressed as a percentage of the total rms voltage, is termed the 'total harmonic distortion' (THD). Audio high fidelity is usually accepted as being below 0.1% THD; purists may demand considerably better than this.

Since distortion invariably is dependent on load impedance, signal level and frequency, it is usually quoted for a power amplifier at several power levels and frequencies with a particular load.

Power amplifier output stage

The requirements of an audio power amplifier are threefold:

1. It should be capable of driving a prescribed power into a loudspeaker load of stated nominal impedance.
2. The voltage gain should be well defined and unaffected by the loudspeaker load and device tolerances.
3. The distortion should be low.

Refer to Horrocks (1990), Chapters 2 and 4.

Criteria (2) and (3) indicate that overall negative feedback should be used. The closed-loop gain will then be determined by the ratio of resistor values and also the output resistance, and the distortion figure will be substantially reduced when feedback is applied.

Class A and class B operation

Goodge (1990), Chapter 8, provides extra useful detail.

The amplifier must be able to drive the peak load current I_L through the load in both directions and, since I_L is likely to be in the order of several amps, such a drive is not within the capability of low-power amplifiers as described in previous chapters. Current gain must therefore be provided in the circuit and the output stage is often realized by using a pair of complementary emitter-followers (Fig. 8.3) which also provide a low output resistance.

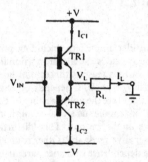

Fig. 8.3 Complementary emitter-follower output stage with dual power supply.

156

The n-p-n emitter-follower TR1 conducts for positive signal excursions while TR2 conducts for negative. This arrangement in which TR1 and TR2 are OFF under no-signal conditions and conduct on alternate half-cycles is referred to as class B 'push-pull' operation. In class A, the transistors are biased at a quiescent current greater or equal to half of the peak signal current, so that they conduct all the time. (It is assumed that the load current is provided symmetrically by both transistors.) Clearly class A is less efficient than class B, since under no-signal conditions there will be a steady power dissipation which will be very significant if high currents are involved. At this stage it is useful to calculate the output-stage performance in terms of power drawn from the supply, power dissipated in the transistors and in the load, as well as the circuit efficiency (defined in Exercise 8.3). This is done now for class B; the reader should perform the analysis for class A.

TR1 and TR2 are matched, i.e. they have equal βs.

All the amplifiers in the previous chapters are examples of class A operation.

For the complementary emitter-follower output stage shown in Fig. 8.3, calculate (a) the power drawn from the supply, (b) the power dissipated in the load, (c) the power dissipated in each transistor, and (d) the efficiency of the circuit. Assume ideal transistors, i.e. they are ON when $V_{BE} \geqslant 0$ V, and a sinusoidal input signal of peak amplitude aV $(0 \leqslant a \leqslant 1)$, where V is the voltage of each power supply.

Worked Example 8.2

Solution. Figure 8.4 shows the currents and voltages in the circuit.
(a) Consider the $+V$ power supply which provides I_{C1}, the load current during the positive half-cycle.

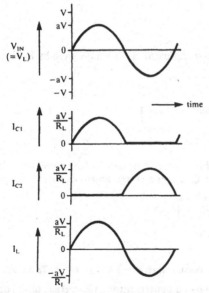

Fig. 8.4 Currents and voltages in the class B circuit of Fig. 8.3 assuming a sinusoidal input and ideal devices.

The instantaneous load current is given by

$$I_L = \frac{aV}{R_L} \sin \omega t$$

and the average I_{C1} over the positive half-cycle ($T/2 = \pi/\omega$) is

$$I_{C1(av)+} = \frac{\omega}{\pi} \int_0^{\pi/\omega} \frac{aV}{R_L} \sin \omega t \, dt$$

$$= \frac{\omega aV}{\pi R_L} \left[-\frac{\cos \omega t}{\omega} \right]_0^{\pi/\omega}$$

$$= \frac{2aV}{\pi R_L}$$

Since I_{C1} is zero during the negative half-cycle, the average I_{C1} over each complete cycle is

$$I_{C1(av)} = \frac{aV}{\pi R_L}$$

Similarly, $I_{C2(av)} = \frac{aV}{\pi R_L}$

The average power delivered by each power supply is the product of voltage and average current and, since there are two supplies, the total average power drawn from the supply is

$$P_{DC} = \frac{2aV^2}{\pi R_L}$$

(b) The average power dissipated in the load is given by Equation 8.1:

$$P_o = \frac{a^2 V^2}{2R_L}$$

which has a maximum when $a = 1$

$$P_{0(max)} = \frac{V^2}{2R_L}$$

(c) The efficiency of the circuit is the ratio of load power to the power provided by the supply (each averaged over a cycle), usually expressed as a percentage.

$$\text{Efficiency} = \frac{a^2 V^2}{2R_L} \frac{\pi R_L}{2aV^2}$$

$$= \frac{a\pi}{4}$$

Maximum efficiency occurs when $a = 1$ and is $\pi/4$ or 78.5%.

(d) Since there are two output transistors, the average dissipation, $P_{C(av)}$, in each transistor must be half of the difference between the supply power and the load power, i.e.

$$P_{C(av)} = \frac{1}{2}(P_{DC} - P_o)$$

$$= \frac{1}{2}\left(\frac{2aV^2}{\pi R_L} - \frac{a^2V^2}{2R_L}\right)$$

$$= \frac{V^2}{2\pi R_L}\left(2a - \frac{a^2\pi}{2}\right)$$

$P_{C(av)}$ must be differentiated with respect to a to determine the maximum device dissipation.

$$\frac{dP_{C(av)}}{da} = \frac{V^2}{2\pi R_L}(2 - a\pi) = 0 \quad \text{for maximum}$$

i.e. $a = 2/\pi$ or 0.637

Hence the maximum device dissipation occurs when the load current is at 63.7% of its maximum and is given by

$$P_{C(av)max} = \frac{V^2}{\pi^2 R_L}$$

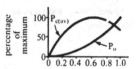

Repeat Worked Example 8.2 for class A operation where the quiescent collector current of the transistors is $V/(2R_L)$. **Exercise 8.2**

[*Answer*: (a) $P_{DC} = V^2/R_L$ (b) $P_{O(max)} = V^2/2R_L$ (c) max efficiency = 50% (d) $P_{C(av)max} = V^2/2R_L$]

Let us return to the class B output circuit. A dual power supply ($+V$ and $-V$) as shown in Fig. 8.3 is not necessary and may usefully be replaced by a single supply ($V_{CC} = 2V$) if the amplifier output is capacitively coupled to the loudspeaker load as shown in Fig. 8.5. The coupling capacitor, C_L, and the load may be connected either way round, as shown, but the arrangement of Fig. 8.5b, with one terminal of the load at earth, is the more common. Naturally, the capacitive coupling curtails the low-frequency response:

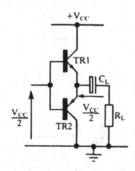

Fig. 8.5 Output stage with single supply.

$$\text{low frequency } -3 \text{ dB point} = 1/(2\pi R_L C_L) \tag{8.3}$$

assuming that the output resistance of each emitter-follower is very much less than R_L.

The results of Worked Example 8.2 and Exercise 8.2 have been converted into single-supply form and are tabulated (Table 8.1) in a useful summary of the performance measures of class A and class B. This helps with device and heatsink specification.

The output emitter-follower provides a useful current gain of approximately β ($I_E/I_B = 1 + \beta$), so the peak base current drive for TR1 and TR2 is I_L/β. As we find in the case study later in this chapter, a current reduction by a factor of β may be insufficient to reduce the required driving signal currents to the milliamp levels of low-power stages and further complementary emitter-followers TR3 and TR4 are usually added to achieve this by a further β reduction (see Fig. 8.6). Compound Darlington pairs TR3/TR1 and TR4/TR2 are immediately recognizable.

Under no-signal (quiescent) conditions, the emitters of TR1 and TR2 will normally be biased mid-way between V_{CC} and earth to permit equal positive and negative voltage swings before limiting.

The significant measure of comparison between class A and B is not the efficiency (50:75) but the transistor dissipation (4:93:1).

TR3 and TR4 are also matched.

See Chapter 4.

The amplified diode

Assume, with reference to Fig. 8.6, that V_{IN} and V_L are at the desired $V_{CC}/2$ quiescent level and then a positive-going signal is superimposed on V_{IN}. Since TR3 and TR1 do not conduct substantially until each base-emitter voltage is in the order of 0.7 V, the output remains at its quiescent level until V_{IN} is approximately 1.4 V above $V_{CC}/2$, at which point the output V_L begins to 'follow' V_{IN}, but 1.4 V behind. Similarly, on negative-going inputs there is a dead zone, delaying conduction in TR4 and TR2.

Table 8.1

	Class A	Class B
Average input power from power supply, P_{DC}(W)	$\dfrac{V_{CC}^2}{4R_L}$	$\dfrac{V_{CC}^2}{2\pi R_L}$ *
Maximum average output power $P_{0(max)}$(W)	$\dfrac{V_{CC}^2}{8R_L}$	$\dfrac{V_{CC}^2}{8R_L}$ *
Maximum efficiency (%)	50	$\dfrac{\pi}{4} \times 100 = 78.5$ *
Maximum average power dissipated in each transistor $P_{C(max)}$(W)	$\dfrac{V_{CC}^2}{8R_L} \left(= P_{0(max)} \right)$ ‡	$\dfrac{V_{CC}^2}{4\pi^2 R_L} \left(= \dfrac{P_{0(max)}}{4.93} \right)$ †

* At maximum voltage swing ($= V_{CC}$)
† At 63.7% of maximum voltage swing
‡ Under no-signal conditions

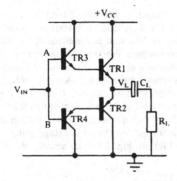

Fig. 8.6   Complementary Darlington emitter-followers.

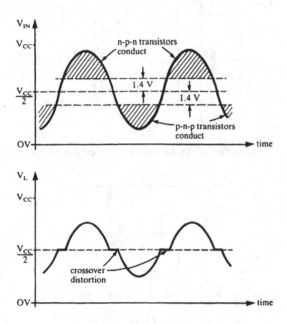

Fig. 8.7   Crossover distortion.

Figure 8.7 illustrates this 'crossover distortion' which is very audible, especially for the lowest-level signals. Crossover distortion can be reduced considerably by providing a small quiescent bias current in TR1 and TR2 so that the direction of the load current may be changed in a smooth manner without discontinuity due to transistor nonlinearity and its attendant distortion. This is called class AB operation. I_Q is typically around 20 mA and is established by a bias voltage applied between

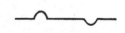

This is called 'thermal tracking'.

points A and B (Fig. 8.6) equal to the sum of the four base-emitter voltages. Each V_{BE} is approximately 0.7 V but, due to transistor spreads, it is impossible to calculate the exact value of V_{AB} that is required. Also since V_{BE} is temperature-dependent, V_{AB} must be a similar function of temperature to maintain a constant quiescent value of I_Q. A crude approach to a solution would be to connect several diodes (two or three) in series between points A and B, noting that the collector current of the preceding stage will flow through them. There are problems with this method: only an integer number of diode voltages (with tolerances) can be used, and it is difficult to ensure that the diodes are at the same temperature as the output transistors. The ideal circuit for biasing the output stage is the amplified diode (or V_{BE} multiplier) shown as TR5 in Fig. 8.8 and already described and analysed in Chapter 5.

$$V_{AB} = (1 + R_2/R_1) V_{BE5} \tag{8.4}$$

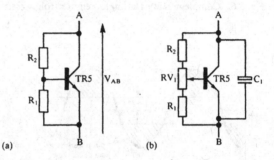

Fig. 8.8 Amplified diode: (a) basic circuit. (b) Adjustable version.

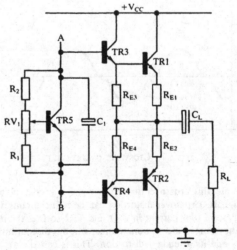

Fig. 8.9 Complete biased output stage.

Note that V_{AB} has the same temperature coefficient as the V_{BE}s of transistors TR1 to TR4 and, if all five transistors are kept at the same temperature, it is possible to adjust V_{AB} (by making the ratio R_2/R_1 adjustable – RV_1 in Fig. 8.8b) and set the output stage current to the desired level. TR5 is usually mounted on the heatsink with TR1 and TR2 (and between them) to ensure that these transistors are at the same temperature. A capacitor C_1 may be added to bypass the amplified diode for a.c. signals (slope resistance is given by Equation 5.35) so that the amplified diode is used only to provide a d.c. offset voltage between its two terminals.

To aid temperature stability of the quiescent bias current at high temperatures where I_C would tend to increase considerably, series feedback is introduced in the form of resistors R_{E1} and R_{E2} in the emitters of TR1 and TR2 (see Fig. 8.9). R_{E1} and R_{E2} are low-valued, typically 0.5 Ω or less. They should be significantly less than R_L otherwise more power would be dissipated in these resistors rather than usefully in the load, reducing both the amplifier power-handling capability and efficiency.

Further resistors R_{E3} and R_{E4} at the emitters of TR3 and TR4 establish in TR3 and TR4 a quiescent current over and above the level I_Q/β. This improves the amplifier linearity and reduces distortion. (By reducing the percentage variation of the current, r_e varies less.)

Voltage gain stages and overall configuration

The output stage comprising the complementary compound emitter-followers and amplified diode provides only current gain; the voltage gain is just less than unity. It is therefore necessary to precede this with at least one stage of voltage gain to

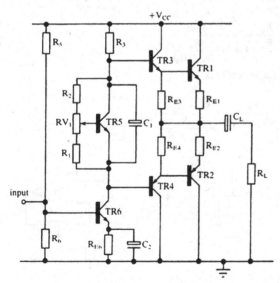

Fig. 8.10 Power amplifier including class A common-emitter stage providing voltage gain.

reduce the signal amplitude required from the preamplifier – the simplest, a single common-emitter stage (TR6), is shown in Fig. 8.10.

Allowance should be made for biasing TR6.

TR6 must be biased at a d.c. quiescent current several milliamps greater than the peak a.c. drive requirement for the output emitter-followers in order that full output may be achieved without limiting. Potentiometer bias is provided by resistors R_5 and R_6 together with the emitter resistor R_{E6} which is decoupled by C_2. The load resistor for this stage is R_3 and must drop, when passing the quiescent collector current of TR6, a voltage approximately equal to $V_{CC}/2$ to enable the d.c. output voltage to be at half the supply rail voltage. Under this bias condition the output can swing by equal voltages in positive and negative directions before being limited by the supply rail or earth.

The voltage gain of the power amplifier configuration shown in Fig. 8.10 is given by the ratio of the total collector load of TR6 to r_{e6}. The total collector load of TR6 is approximately $R_3 \| \beta^2 R_L$ where β^2 is the product of the βs (not necessarily equal) of TR1,3 or TR2,4. Clearly, since this is subject to variations in the device βs, so too is the voltage gain, even if r_{e6} is well defined in a stable bias design. Further, r_{e6} is not constant and its variation with instantaneous signal current level introduces distortion. As noted earlier in this chapter, overall negative feedback should be incorporated to define the gain accurately and to reduce distortion. Shunt or series negative feedback may be used and is shown in its simple operational amplifier (OPA) form in Fig. 8.11.

Refer to Horrocks (1990), Chapter 6.

The closed-loop voltage gain, A_V, of the amplifier incorporating negative feedback is determined by the amplifier's open-loop gain, A_o, and the ratio of the two external resistors, R_A and R_B.

For shunt negative feedback

$$A_V = -\frac{R_B}{R_A} \frac{A_o \dfrac{R_A}{R_A + R_B}}{1 + A_o \dfrac{R_A}{R_A + R_B}} \tag{8.5}$$

For series negative feedback

$$A_V = \frac{R_A + R_B}{R_A} \frac{A_o \dfrac{R_A}{R_A + R_B}}{1 + A_o \dfrac{R_A}{R_A + R_B}} \tag{8.6}$$

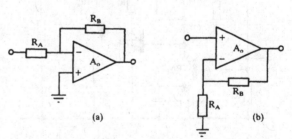

Fig. 8.11 Negative feedback applied to an OPA: (a) shunt feedback, (b) series feedback.

164

Collector load bootstrapping

It is evident from Equations 8.5 and 8.6 that the higher the magnitude of the open-loop voltage gain, A_o, the more closely the closed-loop gain is specified by the resistors alone (lower distortion).

A useful technique which may be used to increase open-loop gain is called 'collector load bootstrapping'. This is another application of Miller's theorem which was the basis of the bootstrap bias circuit described in Chapter 4. The principle here is the same.

Consider the common-emitter amplifier TR6 of Fig. 8.12. Note that the emitter-follower TR1 is connected to the collector of TR6 and its output is capacitively coupled via C_4 back to TR6's collector load resistor which has been split into R_3 and R_4.

If the signal voltage at the collector of TR6 is v, then Av is present at the emitter of TR1 and also at the junction of R_3 and R_4, provided that the impedance of C_4 is very low at the frequencies of interest. A is the voltage gain of the emitter-follower and is given by $A = R/(r_{e1} + R)$ where $R = R_E \| R_3$ and $r_{e1} = 0.025/I_{C1}$. Now the signal voltage across R_4 is $(1 - A)v$. Therefore, for a.c. signals, R_4 appears to have the value $R'_4 = R_4/(1 - A)$ which is very much greater than R_4 since A is close to, and just less than, unity. Hence the effective collector load resistance is significantly increased and so is the voltage gain of the common-emitter amplifier. This is now given by

$$A_V = - \left[R'_4 \| r_{in(follower)} \right] / r_{e6} \quad \text{where} \quad r_{in(follower)} = \beta_1 R$$

and may be compared with the voltage gain without bootstrapping which is

$$A_V = - \left[(R_3 + R_4) \| r_{in(follower)} \right] / r_{e6}$$

Consider the collector load bootstrap circuit of Fig. 8.12 with $R_3 = R_4 = 1.5$ kΩ and $R_E = 3.3$ kΩ. Both TR6 and TR1 are biased at collector currents of 2.5 mA and have βs of 100.

Worked Example 8.3

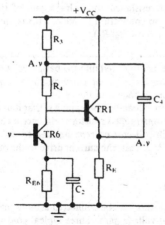

Fig. 8.12 Collector load bootstrap circuit.

(a) With the bootstrapping inoperative (C_4 disconnected), calculate the voltage gain between the base of TR6 and the emitter of TR1.

(b) Recalculate this voltage gain when C_4 (high-valued) is connected.

Solution

(a) $A_V = \dfrac{R_{C6}\|r_{in1}}{r_{e6}} \dfrac{R_E}{R_E + r_{e1}}$

$ = -\dfrac{3k\|330k}{10} \dfrac{3k3}{3k3 + 10}$

$ = -296$

(b) $R_{C6(eff)} = \dfrac{1k5}{1 - A}$

 where $A = \dfrac{R}{R + r_{e1}}$

 and $R = 3k3\|1k5 = 1.03125$ kΩ

$ \therefore A = \dfrac{1.03125}{1.04125} = 0.99$

giving $R_{C6(eff)} = 156$ kΩ

$ \therefore A_V = -\dfrac{156k\|103k}{10} \times 0.99$

$ = -6142$

an increase of over 20 times.

Note that both the input resistance and voltage gain of the emitter-follower are reduced when the bootstrap connection is made. This is due to the lower effective value of emitter resistance (now $R_E\|R_3$).

The bootstrapping can be incorporated into the basic circuit of Fig. 8.10 by splitting R_3 into R_3 and R_4 and connecting their junction to the amplifier output via C_4. (Refer to the complete circuit shown in Fig. 8.14.)

Another aspect of collector load bootstrapping is that it maintains a virtually constant voltage across the bootstrapped resistor and hence a constant current through it. Any (signal) changes in collector current are then forced into the following circuit. In the power amplifier circuit, the current drive to the output emitter-followers is improved in this way.

Overall configuration

Negative feedback components may be connected around the high-gain amplifier to specify closely the overall voltage gain. Three typical configurations are shown in Fig. 8.13 where the compound complementary emitter-followers are represented by

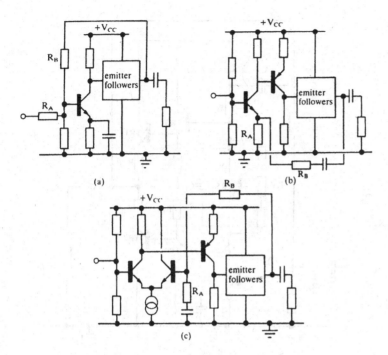

Fig. 8.13 Three typical feedback configurations.

a block and collector load bootstrapping has been omitted.

Figure 8.13a shows a basic shunt voltage feedback (inverting) amplifier with low output resistance and an input resistance approximately equal to the value of the series input resistor. In Fig. 8.13b and c, an additional stage is added to increase the open-loop voltage gain. Series feedback is applied to these non-inverting circuits which can have higher input resistance (dependent on the bias resistors) than the other circuit.

A detailed circuit of the shunt feedback configuration is given in Fig. 8.14. In Fig. 8.10, the bias resistor, R_5, was shown returned to the positive supply rail. If, instead, it is connected between the base of TR6 and the output point (junction of R_{E1} and R_{E2}), its function doubles as a shunt feedback resistor equivalent to R_B in Fig. 8.11a. Addition of R_7 (R_A) and the input coupling capacitor, C_3, completes the circuit.

An audio power amplifier design procedure

Specification

Average output power = 10 W into 4 Ω
Input sensitivity   < 0.5 V (rms) for full output
Input resistance   > 1 kΩ
Circuit configuration to be that shown in Fig. 8.14

Input sensitivity is defined as the input signal amplitude which will generate the rated output.

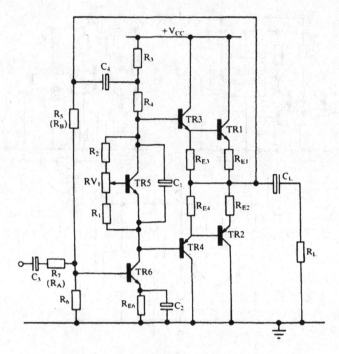

Fig. 8.14 Complete power amplifier circuit.

Supply voltage requirements

Equation 8.2 is relevant:

$$P_O(\text{max}) = V_{CC}^2/8R_L$$

assuming that the output can swing through the full V_{CC} supply.

i.e. $10 = V_{CC}^2/32$

$\therefore\quad V_{CC} = \sqrt{320} = 18 \text{ V (approx)}$

In practice, the output cannot swing through the full V_{CC} range due to V_{BE} drops, the potential divider action of R_{E12} and R_L, and driver stage limitations. V_{CC} must therefore be greater than 18 V and, from experience, a 25 V supply would prove a suitable minimum figure. This will be checked later in the design.

Output stage considerations

Output transistors
Select transistors TR1/TR2 in Fig. 8.14 which can handle the peak voltage and current requirements.

$$V_{CE}(\text{peak}) = V_{CC}(\text{approx})$$

$$I_c(\text{peak}) = V_{CC}/2R_L = 25/8 = 3.125 \text{ A}$$
$$(\text{more exactly, } 18/8 = 2.25 \text{ A})$$

The TIP 31A/32A complementary pair satisfies these criteria with ratings of 60 V and 3 A. The maximum dissipation in each transistor is 2 W, approximately one fifth of the maximum output power (see Table 8.1). These devices are rated at 2 W at 25 °C free-air temperature without a heatsink, so a small heatsink will be quite adequate at higher ambient temperatures. (They are capable of dissipating 40 W at a junction temperature of 150 °C, but this would require a very large and efficient heatsink.)

Calculation of heatsink requirements is covered in Sangwine (1987), Chapter 6.

Drive to TR1/TR2

> Peak emitter current $= 2.25$ A
> Current gain, β $= 25$ (minimum) for TIP31A/32A
> ∴ Peak base current $= 90$ mA

Drive to TR3/TR4

Select TR3/TR4 to satisfy peak voltage and current requirements – the BC 182/212 complementary pair is suitable (50 V, 100 mA). Their minimum β of 80 limits the peak drive requirement at their bases to a maximum of 1.125 mA. The quiescent collector current of TR6 must therefore be greater than this.

Output stage emitter resistors

R_{E1} and R_{E2} must be much less than R_L (4 Ω) but should be as high-valued as possible to achieve maximum bias stability.

$R_{E1} = R_{E2} = 0.47$ Ω is a satisfactory compromise. R_{E3} and R_{E4} may be 150 Ω, establishing in TR3 and TR4 a quiescent current of 5 mA over and above the quiescent base currents of TR1 and TR2.

Amplified diode (TR5)

With the output transistors (TR1 and TR2) conducting at a quiescent level of 20 mA (to reduce crossover distortion), the voltage drop across each emitter resistor R_{E1} and R_{E2} is only $0.02 \times 0.47 = 10$ mV, which is negligible compared with the V_{BE}s of the four emitter-followers. The voltage across the amplified diode (BC 182) should then be in the order of 2.8 V (4 × 0.7 V) and should be adjustable around that value to accommodate device spreads.

$RV_1 = 1$ kΩ, $R_1 = 1$ kΩ and $R_2 = 3.3$ kΩ are satisfactory component values (range of 2.65 to 4.3 times V_{BE5}).

C_1, which bypasses the amplified diode to a.c. signals, may be omitted if the slope resistance of the amplified diode is very much less than the resistance of the bootstrapped collector load of TR6 (see later).

TR5 should be mounted on the same heatsink as the output power transistors TR1 and TR2, and between them to achieve good thermal coupling.

Driver stage

With a 25 V power supply, the quiescent output voltage should be approximately 14 V for maximum signal handling since allowance must be made for V_{CE6} and the voltage across R_{E6}.

$$\therefore V_{B3} = 14 + (2 \times 0.7) \text{ V}$$
$$= 15.4 \text{ V}$$

The voltage across R_3 and R_4 is

$$(V_{CC} - V_{B3}) = 25 - 15.4 \text{ V}$$
$$= 9.6 \text{ V}$$
$$= I_{C5}(R_3 + R_4)$$

$\therefore$ with $I_{C6} = I_{C5} = 3$ mA, $R_3 + R_4 = 3.3$ kΩ

Select $R_3 = 1.8$ kΩ, $R_4 = 1.5$ kΩ, an approximately equal split of $R_3 + R_4$. If simulation (e.g. using SPICE) shows this to be unsatisfactory, other values should be tried. Also, selecting the value of the bootstrap capacitor, C_4, as 80 μF should be tested by simulation or by measurement of the frequency response of the constructed circuit.

With I_{C5} equal to 3 mA and using Equation 5.35, it is now possible to calculate the slope resistance of the amplified diode as approximately 67 Ω. Since this is obviously much less than the collector load of TR6, the signal voltages at the inputs of both the n-p-n and the p-n-p emitter-followers are almost equal without requiring C_1 as an a.c. bypass.

For highly stable bias, approximately 2 to 3 V should be dropped across R_{E6} but this would reduce the available output swing or increase the required power supply voltage. However, the overall negative feedback helps to stabilize the bias and a compromise figure of 0.5 V is used. With $I_{E6} = I_{C6} = 3$ mA, $R_{E6} = 180$ Ω. The decoupling capacitor C_2 is chosen at 100 μF, giving an associated open-loop break frequency of approximately 190 Hz.

Using Equation 4.16.

Now,

$$V_{B6} = (I_{E6}R_{E6} + V_{BE}) = 0.54 + 0.7 = 1.24 \text{ V}$$

and

$$I_{B6}(\text{max}) = 3/80 \text{ mA} = 37.5 \ \mu\text{A}$$

If R_5 is chosen as 100 kΩ

$$V_{R5} = 14 - 1.24 \text{ V}$$
$$= 12.76 \text{ V}$$

and $I_{R5} = 12.76/100$ mA
$$= 127.6 \ \mu\text{A}$$

$$\therefore I_{R6} = I_{R5} - I_{B6}$$
$$= 90 \ \mu\text{A (approx)}$$

$$\therefore R_6 = 1.24/0.09 \text{ k}\Omega$$
$$= 15 \text{ k}\Omega \text{ (nearest preferred value)}$$

At this stage it is possible to check that the supply voltage chosen (25 V) is adequate for the required 18 V peak-to-peak load voltage swing plus other voltage drops in the circuit.

$$V_{CC} \geq V_L(\text{pk-pk}) + I_O(\text{pk})(R_{E1} + R_{E2}) + I_{E6}R_{E6} + V_{BE6} + V_{BE1} + V_{BE3} + V_{BE2} + V_{BE4}$$

$$\geq 18 + (2.25 \times 0.94) + 0.54 + 0.7 + 0.7 + 0.7 + 0.7 + 0.7$$

$$\geq 24 \text{ V approx.}$$

This calculation shows that there is some voltage in hand and the output power specification should be met.

Input sensitivity and voltage gain

The input sensitivity has been specified as less than 0.5 V rms for full 10 W output.

The rms output voltage at 10 W is given by

$$V_o^2(\text{rms}) = R_L P_o$$
$$= 4 \times 10 = 40$$

$$\therefore V_o(\text{rms}) = 6.3 \text{ V}$$

Assuming a high open-loop voltage gain, the closed-loop voltage gain of the amplifier is given by

$$A_V = -R_B/R_A = -R_5/R_7 = -6.3/0.5 = -12.6$$

Now, since $R_5 = 100 \text{ k}\Omega$, $R_7 = 100/12.6 \text{ k}\Omega = 7.9 \text{ k}\Omega$ (maximum). If R_7 is specified as 4.7 kΩ, attenuation introduced by volume control circuitry may be accommodated.

Low-frequency response

The frequency response at low frequency (LF) is dominated by the output coupling capacitor C_L selected (for cost reasons) at 1000 μF. This value, in conjunction with the 4 Ω load, gives a LF breakpoint at $f_o = 1/2\pi \times 4 \times 10^3 = 40$ Hz. Increasing C_L to 2000 μF would give 20 Hz.

The input coupling capacitor C_3 should be chosen so that the breakpoint due to C_3 is at a much lower frequency than 40 Hz. $C_3 = 100$ μF is more than adequate.

Comments

The design is a compromise between d.c. bias stability, voltage gain and input resistance.

The input resistance is given by R_7 (4.7 kΩ). To improve the bias stability the standing bias current through R_5 and R_6 would have to be raised. With a correspondingly lower value of R_5, R_7 must also be reduced to preserve the voltage gain at the original figure. The input resistance is then reduced, causing a greater loading on the preamplifier.

This is an example of phase lag compensation used to achieve freedom from oscillation. See Bissell (1988), Chapter 8.

In order to suppress unwanted high-frequency oscillations, it is found necessary to connect a 22 pF capacitor between collector and base of TR6.

The transient response under inductive load conditions may be improved by incorporating a series resistor and capacitor (4.7 Ω + 0.1 μF) in parallel with the load. This is called a Zobel network.

This network absorbs potentially high-amplitude back EMFs which may be generated transiently (by the loudspeaker inductance) and could damage the output transistors.

The design has now been carried out and the circuit is shown complete with component values in Fig. 8.15.

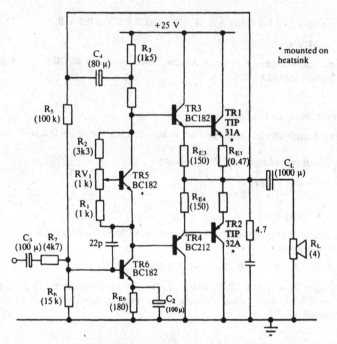

Fig. 8.15 Complete power amplifier design.

Performance

A power amplifier was built to the above design and performed as follows:

Output power = 10 W into 4 Ω
Input sensitivity = 0.5 V rms for full output
−3 dB bandwidth = 40 Hz to 60 kHz
Distortion = 0.3% THD at full rated output into 4 Ω

The low-frequency response is dominated by C_L as predicted. Distortion is rather high at full output; this is due to nonlinearity in the common-emitter stage TR6 and could be reduced by introducing local series feedback in this stage, compensating for the reduced open-loop gain by adding an extra feedback stage as in Fig. 8.13.

It is interesting to look at the measured open-loop and closed-loop frequency responses (Fig. 8.16). The high open-loop gain is achieved by collector load bootstrapping but the high-frequency breakpoint at 1.8 kHz is very low. However, applying overall negative feedback, as well as reducing the gain, extends the break frequency in direct proportion. Similarly, the LF effect of the capacitors within the feedback loop (C_2 and C_4) is improved by the feedback.

Exercise 8.3 Predict the open-loop voltage gain of the power amplifier shown in Fig. 8.15 with and without collector load bootstrapping. Assume that the product of the βs of

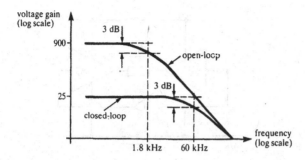

Fig. 8.16 Measured frequency response of the amplifier.

TR1/TR3 (and TR2/TR4) is 4800.
[*Answer*: With bootstrapping, $A_V = -1000$; and without bootstrapping, $A_V = -350$. The reasons for the increase in $|A_V|$ not being as great as might be expected are the potential divider action of $R_{E1,2}$ and R_L reducing the follower gain, and also the shunting of R_4(effective) by the input resistance $\beta^2(R_L + R_{E1,2})$ of the follower.]

Other aspects of power amplifiers

Within this chapter it is impossible to present an exhaustive coverage of audio power amplifiers. While the fundamentals have been treated from a design viewpoint, there are several additional important aspects, rather dissociated, which this section attempts to cover.

Bridge configuration

It has been evident throughout this chapter and clearly expressed in Equation 8.2 that, for a given loudspeaker impedance, the output of a power amplifier is limited by the magnitude of the power supply voltage. This limitation is important in mobile situations where the vehicle's battery is the source of power. Indeed, using a 4 Ω loudspeaker and assuming a full signal voltage swing of 12 V, the maximum output power is restricted to only 4.5 W. Lower loudspeaker impedances are uncommon (but an output transformer could be used) and it is impractical to install additional batteries in series with the main one, so this output power appears to be the limit. However, there exists a technique for effectively doubling the voltage swing by using two power amplifiers (per channel) in an antiphase or 'bridge' arrangement as shown in Fig. 8.17.

A_1 and A_2 are identical power amplifiers operating with a supply voltage of magnitude V_{CC}. The input voltage, v_{in}, is connected directly to A_1 and through an inverter to A_2 so that the voltage signals at the outputs of A_1 and A_2 are equal in magnitude but of opposite phase. Hence, for ideal amplifiers, the maximum peak-

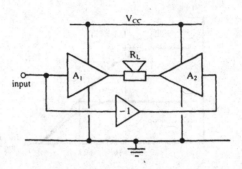

Fig. 8.17 Bridge output configuration.

to-peak voltage across the loudspeaker is $2V_{CC}$ and the maximum output power is now given by

$$P_{O(\text{max})} = (2V_{CC})^2/8R_L = V_{CC}^2/2R_L$$

For $V_{CC} = 12$ V and $R_L = 4$ Ω, the maximum output power is now 18 W. This factor of four times increase in power output (per channel) is a significant benefit for in-car entertainment. To increase the power output even more, it is necessary to use a d.c.-to-d.c. converter to create a higher supply voltage from the vehicle's battery.

I_o doubles and the effective load on each amplifier is 2 Ω.

Use of power MOSTs

Complementary enhancement MOSTs, used as source-followers, are very suitable as output devices in audio power amplifiers. There is no need for driver transistors provided the voltage gain stage can provide the relatively small current needed to charge the MOST junction capacitances.

Note that there is still a need to bias the output stage into conduction to remove crossover distortion. An inter-gate bias greater than the sum of the two threshold voltages must be provided.

An advantage of using MOSTs rather than BJTs in power output stages is their relative freedom from a destructive process called 'thermal runaway'. In a BJT class B amplifier, a temperature rise causes an increase in leakage current which adds to the quiescent current and, if the bias is not adequately stabilized or the heatsink is inadequate, a resulting increase in dissipation raises the device temperature. This reinforcing cycle of events is thermal runaway which can cause BJTs to overheat and fail. FETs are immune to this problem since the channel has a positive temperature coefficient of resistance (at higher drain currents) which tends to stabilize the drain current.

Class D operation

In a quest to maximize efficiency by minimizing device dissipation, it is possible to use the output transistors in the switching mode, driven between the OFF and ON states in both of which, ideally, no power is dissipated. (Except at the instant of switching, either the current through the device or the voltage across the device is zero, as shown in Fig. 1.19.) To implement this mode of operation, which is known

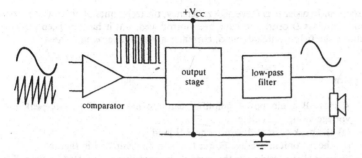

Fig. 8.18 A class D amplifier.

as class D, the audio input signal must be converted into a pulse-width-modulated (PWM) signal. As shown in Fig. 8.18, this may be done using a comparator with the audio signal connected to one input and a high-frequency reference triangular wave (of amplitude greater than the audio signal) connected to the other. The comparator output is a series of pulses whose width corresponds to the amplitude of the audio and drives the output stage to its extremes.

The output stage is connected to the loudspeaker load via a passive low-pass filter to remove the high-frequency components of the PWM signal and recover the relatively low-frequency audio.

In practice the switching states are not perfect – an ON device will have a small but non-zero voltage across it, and an OFF device will exhibit a leakage current. Also, power will be lost in the filter. However, efficiencies of greater than 90% can be achieved using class D operation.

Summary

In this chapter we have introduced the fundamental design aspects of audio power amplifiers, drawing on many of the circuit techniques already featured in this book.

Development of a power amplifier circuit necessarily starts at the output stage with consideration of its voltage and current capabilities dictated by the specified load impedance and power output. Interrelationships of these parameters were analysed at the outset. It has been shown that class B amplifiers have higher power efficiency than class A. Biasing of the output stage is a compromise between efficiency and crossover distortion; usually class AB is favoured and an amplified diode circuit, with its thermal tracking of the output transistors, is ideal for stable bias. With the output emitter-followers providing only current gain, one or more voltage gain stages are required, possibly aided by collector load bootstrapping, to achieve a high open-loop voltage gain. Application of overall negative feedback then produces a closed-loop voltage gain closely defined by a resistor ratio giving low distortion and an extended bandwidth. A detailed design of an audio power amplifier has been presented as a case study.

A bridge output configuration has been shown to increase the output power and the advantage of MOSTs in removing the threat of thermal runaway has been discussed. In order to achieve high efficiency, the techniques of pulse-width modulation and class D operation have been introduced, but it has not been possible to explore the field of digital audio, currently a very active research area.

Problems

8.1 A class B audio power amplifier can provide a 12 V peak-to-peak output voltage swing. Calculate
(a) the peak current drawn by an 8 Ω load
(b) the maximum average power that can be dissipated in that load.
Show that the results are the same if the amplifier is operated in class A.

8.2 An audio power amplifier is required to deliver an average power of 50 W into a loudspeaker of 8 Ω impedance (assumed resistive). Calculate the minimum possible power supply voltage which must be provided.

8.3 Refer to the collector load bootstrap circuit of Fig. 8.12.
$R_3 = R_4 = R_E = 5$ kΩ; $V_{CC} = 40$ V; $I_{C6} = 2$ mA; $\beta_{TR1} = \beta_{TR6} = 150$.
Calculate the voltage gain from the base of TR6 to the emitter of TR1 both with and without bootstrapping.
(Assume that the bootstrap and emitter decoupling capacitors have zero reactances at the frequency of interest.)

8.4 Repeat Problem 8.2 assuming a bridge-configured power amplifier. What is the peak current in the loudspeaker?

Power supply regulators 9

□ To summarize the essential design features of unregulated power supplies.
□ To explain the terms *'regulation'*, *'ripple'* and *'ripple factor'*.
□ To describe in detail the operation of linear voltage regulator circuits and protection circuitry.
□ To design series regulators using BJTs and op-amps.
□ To introduce the basic principles of switching regulators.

Objectives

Introduction

Electronic systems invariably require a source of d.c. power and, except in the case of battery-powered (portable) equipment, the a.c. mains (240 V rms, 50 Hz) must be converted to the appropriate d.c. voltage. This is performed in a power supply unit (PSU) which usually comprises a transformer, rectifier, filter and regulator (Fig. 9.1). The first three components which form what is called an unregulated or 'raw' d.c. supply are not covered in detail in this chapter. Principles of these are outlined to enable component selection. Attention is focused on the operation and design of regulator circuits which call on many of the circuit techniques introduced in earlier chapters.

Sangwine (1987), Chapter 4, and Goodge (1990), Chapter 11, are useful references on power supplies.

The unregulated supply

As we shall see, the unregulated supply is not high quality – its d.c. output voltage can fall quite significantly with load current demand and it has, superimposed on the mean d.c. voltage, an a.c. component (or **hum**) at a mains-related frequency.

Transformer

In a power supply, the transformer is the component which converts (or transforms) the a.c. mains to a higher or (usually) lower a.c. voltage. Fundamentally a transformer consists of two coils (or windings) inductively coupled by a magnetic core. The input winding is called the primary, the output is the secondary and the ratio of primary to secondary voltages is equal to the ratio of the number of turns in each winding.

The mains is often referred to as **line**.

If a transformer primary has 1500 turns with 240 V rms applied and the secondary has 150 turns, calculate

(a) the turns/volt, and
(b) the voltage across the secondary winding.
[*Answer*: (a) Turns/volt = 1500/240 = 6.25 (b) Secondary voltage = 150/6.25 = 24 V rms. (Quite a large number of turns is required to ensure close linkage between the windings.)]

Exercise 9.1

Mains voltage, while nominally 240 V rms, is subject to a ±6% variability.

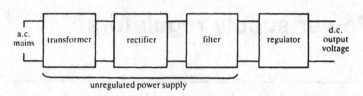

Fig. 9.1 Block diagram of d.c. power supply.

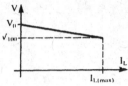

Sometimes the voltage V_{10}, at 10% of maximum load, is used instead of V_0.

The secondary/primary voltage ratio is not sustained as current drawn by a load connected across the secondary is increased. This is due to losses: magnetic (or iron) losses arising from eddy currents in the core, and resistive (or copper) losses caused by the resistance of the wire used in the coils. The fall in voltage due to losses is called **regulation** and is commonly defined as

$$\text{Regulation} = \frac{V_0 - V_{100}}{V_0} \times 100\%$$

where V_0 is the no-load secondary voltage and V_{100} the voltage under full load current. It is usual for a transformer secondary voltage to be specified at the full load current which is calculated from the quoted VA (power) rating and secondary voltage.

A separate transformer is not required for each secondary voltage; several secondaries may be wound on the same core, sharing a common primary.

Rectifier

The waveform of the transformer secondary voltage is bidirectional and nominally sinusoidal. To convert this a.c. voltage to d.c., a device called a rectifier is used. This is usually one or more silicon diodes selected to handle the required voltage and current.

The secondary is used for only half of the time.

Reversing the diode polarity gives a negative half-cycle output.

The simplest rectifier circuit involves a single diode (Fig. 9.2) and is called a half-wave rectifier since the output appears only on alternate half-cycles when the diode conducts. This is relatively inefficient use of the transformer and the mean output voltage (averaged over a whole cycle) is only V_{max}/π, where V_{max} is the peak value of the secondary voltage, assuming that the voltage across the diode when it conducts is zero.

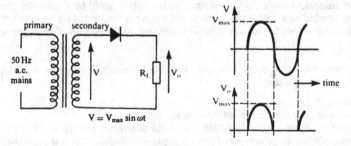

Fig. 9.2 Half-wave rectifier circuit.

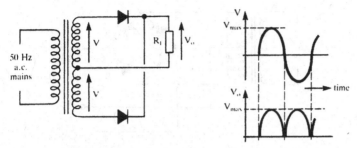

Fig. 9.3   Full-wave rectifier using two secondary windings.

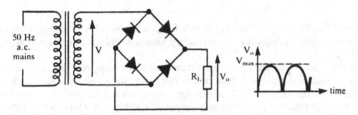

Fig. 9.4   Full-wave bridge rectifier circuit.

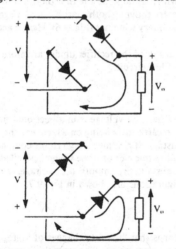

Fig. 9.5   Conduction paths in a bridge rectifier.

A more efficient arrangement is the full-wave rectifier circuit shown in Fig. 9.3. Here, two secondary windings (or a single, centre-tapped winding) provide anti-phase voltages to two diodes which conduct alternately and provide an output voltage during each half-cycle. For a mains frequency of 50 Hz, the output waveform now repeats every 10 ms rather than every 20 ms as in the half-wave

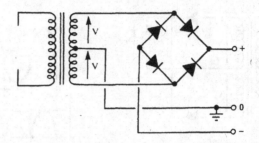

Fig. 9.6 Bridge rectifier providing positive and negative outputs.

rectifier. The mean output voltage is now $2V_{max}/\pi$. However, this arrangement has the disadvantage of requiring two secondary windings, only one of which is used in any half-cycle.

An alternative full-wave rectifier circuit, called a bridge rectifier, uses only a single secondary but requires four diodes (Fig. 9.4). Since pairs of diodes conduct alternately as shown in Fig. 9.5, in practice there are two diode voltage drops in the path between secondary winding and output, the voltage of which is correspondingly lower (by one diode voltage drop) than in the other rectifier circuits.

The bridge rectifier, which usually has the four diodes contained in one package, is useful in being able to provide both positive and negative outputs from a transformer with two secondary windings (Fig. 9.6). Here it acts as two separate full-wave rectifiers.

Voltage and current ratings of the rectifier diodes are more realistically discussed in conjunction with the filter circuit.

Filter

At the output of the rectifier, the voltage is unidirectional, i.e. it is either positive or negative with respect to earth, depending on which way the diodes are connected, but it is by no means constant. It contains a high ripple content at 50 Hz (half-wave) or 100 Hz (full-wave) and harmonics of these frequencies. The ripple can be largely filtered out, leaving a substantially smooth d.c. voltage. Two common filter (or smoothing) circuit configurations are shown in Fig. 9.7.

Capacitor filter

A property of a capacitor is that 'it resists changes of voltage across it' and hence provides the desired smoothing action. The smoothing capacitor is charged from the secondary via the rectifier. When the diode is reverse biased, the capacitor discharges into the load. This is illustrated for a half-wave rectifier in Fig. 9.8.

The periods labelled A, B and C in the figure cover different actions taking place. During A, after switch-on, the capacitor charges through the forward biased rectifier diode in series with the secondary winding resistance, following V up to V_{max}. At the start of B, the diode turns off and the capacitor supplies the load current on an exponential decay with time constant CR_L, or linearly if the load

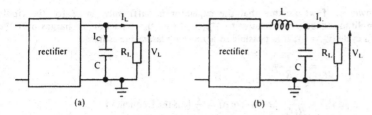

Fig. 9.7 Two common power supply filter circuits: (a) capacitor filter, (b) choke-input filter.

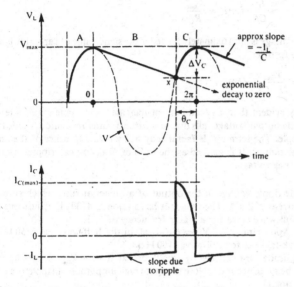

Fig. 9.8 Waveforms for a capacitor filter (half-wave rectifier).

current is constant. With a sufficiently large value of capacitor, the capacitor voltage decays only partially before, in period C, it is recharged to its peak voltage. Periods B and C are repeated. The length of period C is called the **conduction angle**, θ_C, since it is the fraction of the total repetitive period over which the diode conducts.

$$\theta_C = \frac{C}{B + C} \times 2\pi$$

The cyclic change in the output voltage is termed **ripple** and its amplitude obviously increases with increasing load current and a smaller capacitor value. Accurate calculation of ripple amplitude is complex because of the exponential decay.

However, if we assume that the capacitor is sufficiently large for the ripple amplitude to be small relative to V_{max} and that the capacitor discharges over the full cycle ($\theta_C = 0$), it is possible to achieve a reasonable estimate.

$$\frac{dV_C}{dt} = -\frac{V_{max}}{R_L C}$$

$$\therefore |\Delta V_C| = \frac{V_{max}}{R_L C} \Delta t \quad \text{where } \Delta t = \frac{1}{f} \ (f \text{ is the frequency})$$

$$\therefore \text{ peak-to-peak ripple voltage, } \Delta V_C = \frac{V_{max}}{R_L C f} \text{ (half-wave)}$$

$$\approx \frac{I_L}{Cf} \quad \text{for } I_L \approx \frac{V_{max}}{R_L}$$

The ripple voltage is approximately halved if the same value of capacitor is used with a full-wave rectifier.

$$\Delta V_C = \frac{V_{max}}{2 R_L C f} \quad \text{(full-wave)}$$

It is plainly evident that, as load current increases, the ripple amplitude increases and the mean output voltage falls due to the transformer regulation, diode resistance and the ripple. The term **ripple factor** may be used as a measure of the amount of ripple at the output. It is defined as the ratio of peak-to-peak ripple voltage to the average output voltage.

Exercise 9.2 Estimate the ripple voltage at the output of a capacitor filter which supplies 20 V at a load current of 0.5 A. The capacitor has a value of 4700 μF. Consider both half-wave and full-wave cases for a mains frequency of 50 Hz.
[*Answers*: Approximately 2 V peak-to-peak in the half-wave case (50 Hz ripple), and 1 V peak-to-peak for full-wave (100 Hz).
(These amplitudes are sufficiently small compared with V_{max} for the approximate formula to be quite accurate. Note the quite high amplitude of ripple even with this large capacitor.)]

Refer to Goodge (1990), pp. 350–354.

At this stage it is useful to consider expressions for capacitor current, I_C, and diode current, I_D. I_C is a maximum ($I_{C(max)}$) at the start of the conduction period (point X in Fig. 9.8) and it can be shown that

$$I_{C(max)} = 2\pi f C V_O \sin \theta_C \text{ where } \theta_C = \left[\cos^{-1} \left(\frac{V_{max} - \Delta V_C}{V_{max}} \right) \right]$$

V_O is the mean output voltage, assumed equal to V_{max}, and f is the mains frequency (50 Hz). This expression assumes that the charging path has zero resistance. In practice the rectifier diodes and winding resistance reduce $I_{C(max)}$ below this level and also delay the charging so that V_L does not actually reach V_{max} and charging continues beyond the peak of V. This leads to rather odd current waveforms but the high peak current, with its sharp edges, is a source of high-frequency power which can propagate both through the regulator to the load, and back through the transformer into the mains supply where it can cause interference in neighbouring equipment.

The diode current not only charges the capacitor but also supplies the load current when the diode is forward biased. Therefore

$$I_{D(max)} = I_{C(max)} + I_L$$

Estimate the peak currents in the capacitor and the rectifier for the full-wave case in Exercise 9.2 [*Answers*: $I_{C(max)} = 9.2\,\text{A}$, $I_{D(max)} = I_{C(max)} + I_L = 9.7\,\text{A}$.
(These are high values and, in practice, would be somewhat less when secondary winding and diode resistances are taken into account.)]

Exercise 9.3

Choke-input filter

Instead of using only a capacitor to smooth the output voltage, an inductor (or 'choke'), L, may be connected between the rectifier and load, making use of the property of an inductor which 'resists changes of current through it'. This addition reduces current peaks into the capacitor, C. If the smoothing action is regarded as being that of an LC low-pass filter, it would be expected that the mean level of the load voltage is approximately $2V_{max}/\pi$ (assuming full-wave rectification and ignoring diode voltage drops) with an essentially sinusoidal ripple at 100 Hz. This is the case if the values of L and C are sufficiently large.

It can be shown that the peak-to-peak ripple voltage, ΔV_L, is given by

Refer to Goodge (1990), pp. 364–365.

$$\Delta V_L \approx \frac{V_0}{12\pi^2 f^2 LC} \quad \text{where } V_0 = \frac{2}{\pi} V_{max}$$

$$\text{With } f = 50\,\text{Hz}, \Delta V_L \approx \frac{3.38 \times 10^{-6}}{LC}$$

There is a problem at low levels of load current as shown in Fig. 9.9. The regulation is poor (and the ripple is non-sinusoidal).

However, if the load current is increased above a critical level given by

$$I_{L(crit)} = \frac{V_0}{942L}$$

the regulation is dramatically improved, although still degraded by the transformer regulation. This is equivalent to saying that, for a minimum load current $I_{L(min)} = V_0/R_{max}$, there is a critical minimum value of inductance

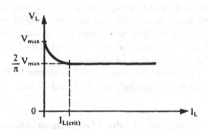

Fig. 9.9 Regulation of choke-input filter.

$$L_{crit} = \frac{R_L}{6\pi f} \text{ or } \frac{R_L}{942}$$

The ripple has a sinusoidal amplitude which, unlike that in a capacitor filter, does not vary substantially with load current.

Worked Example 9.1 A full-wave rectifier output is smoothed by a choke-input *LC* filter. Calculate the values of *L* and *C* required to limit the peak-to-peak output ripple to 1 V superimposed on an average output voltage of 20 V when the load current is 0.5 A.

Solution. From the above equations

$$0.5 = \frac{20}{942L} \text{ giving } L \geq \frac{40}{942} = 0.04 \, \text{H}$$

Use an inductor value of 100 mH (larger value satisfactory).

$$C \geq \frac{3.38}{0.1} \, \mu\text{F} \approx 33 \, \mu\text{F}$$

Use a capacitor value of 47 μF to accommodate tolerance of -20%, $+50\%$. (Compare this capacitor value with that in Exercise 9.2!)

Rectifier and capacitor ratings

In circuit design it is important that components are selected not just by value but by their maximum voltage, current and power ratings as well, otherwise component failure can occur. Rectifier diodes and smoothing capacitors are no exceptions.

Rectifier diodes

In a bridge rectifier the maximum reverse bias on each diode is equal to V_{max}, the peak secondary voltage. For a half-wave rectifier and a full-wave rectifier using two secondary windings, the maximum reverse bias is twice the peak secondary voltage. The peak-inverse voltage (PIV) rating of the rectifier should exceed these figures.

Rectifier diodes have three current ratings indicating their ability to withstand different forms of current load: the mean forward current, the peak repetitive current and the peak non-repetitive current. For full-wave and bridge circuits, the average current per diode is one half of the d.c. output current; in a half-wave rectifier, the single diode carries the full current. The peak repetitive current is the peak current required every cycle or half-cycle to charge the smoothing capacitor *and* provide the load current. A worst-case formula for this has been quoted, assuming negligible winding and diode resistances. The non-repetitive current surge occurs when the supply is switched on and, in a capacitor filter, may be as high as the peak secondary voltage divided by winding plus diode resistance.

Refer to Goodge (1990), p. 365.

It should be noted that a power rectifier diode has a higher voltage drop compared with a signal diode and power dissipation may be important.

A resistor is often added in series with the secondary winding to reduce the peak current flowing and hence relax the ratings of the rectifier and capacitor.

Capacitor

Apart from calculating the required value of capacitance, selection of a smoothing capacitor must also ensure that its voltage and current ratings are not exceeded. These

184

are maximum d.c. voltage (V_{max} or $\sqrt{2}$ times the rms secondary voltage) and rms ripple current.

In estimating capacitor ripple current in a capacitor filter, some designers advocate regarding the ripple voltage as sinusoidal with rms amplitude equal to $\Delta V/(2\sqrt{2})$ and frequency, f_r, equal to the ripple frequency. Capacitor ripple current is then $2\pi f_r C\Delta V/(2\sqrt{2})$. This can be a serious underestimate since the repetitive surge of charging current in a capacitor filter is high. To estimate accurately the rms ripple current it is necessary to refer to graphs in specialist texts. However, it should be borne in mind that, particularly for high load currents, the filter should be redesigned as a choke-input filter which requires a much lower ripple current rating for its capacitor. This may be calculated from

Refer to Texas Instruments (1989) or to Schade (1943).

$$\text{rms ripple current} = \frac{V_0}{1333L}$$

Linear voltage regulators

An unregulated d.c. power supply may be adequate for some applications but there is often a need for the supply voltage to be constant (good regulation) as well as having a low amplitude of ripple. A circuit, called a voltage regulator or stabilizer, connected between the unregulated supply and the load performs this function (Fig. 9.10).

Current regulators are not considered here.

While the specification of power supplies is usually in terms of regulation and ripple amplitude over a defined range of load currents, the measures of performance of a regulator circuit are analogous but somewhat different. The usual regulator parameters are output resistance (r_0) which corresponds to regulation, and ripple-reduction factor (RRF) which quantifies the action of the regulator in reducing the ripple amplitude at its output. They are defined with reference to the following equation which states that changes in output voltage, ΔV_0, are related to changes in input voltage, ΔV_s, and changes in load current, ΔI_0.

or 'ripple-reduction ratio',' regulation factor', etc.

$$\Delta V_0 = \text{RRF}\,\Delta V_s - r_0\,\Delta I_0$$

Hence, $\text{RRF} = \dfrac{\Delta V_0}{\Delta V_s}\bigg|_{\Delta I_0 = 0}$ and $r_0 = -\dfrac{\Delta V_0}{\Delta I_0}\bigg|_{\Delta V_s = 0}$

ignoring the effects of temperature variations

(The minus signs in the above expressions might cause some confusion. They arise from the direction of I_0 labelled in Fig. 9.10. An increase of I_0 (ΔI_0 positive) causes a fall in V_0 (ΔV_0 negative). Since the output resistance r_0 is positive, there must be a minus sign associated with r_0 in the equation.)

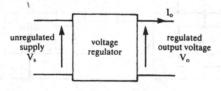

Fig. 9.10 Regulated power supply.

As well as considering large changes of voltage and current (ΔV_S, ΔV_O and ΔI_O), the small-signal equivalents are also valid (v_s, v_o and i_o).

The terms **input (or line) regulation** and **output regulation** are often met in manufacturers' data; they are defined as follows. Input regulation is the change in output voltage for a given change in input voltage (e.g. 7 V to 25 V) and may be expressed as a percentage of the output voltage. Output regulation is the change in output voltage for a given change in load current (e.g. 5 mA to 1.5 A) and may be expressed as a percentage of the output voltage. Also the term **ripple rejection** is used, expressed in dB. This is related to RRF by:

$$\text{ripple rejection} = 20 \log_{10}\left(\frac{1}{\text{RRF}}\right) \text{ dB (at ripple frequency)}$$

There are two basic types of voltage regulator – linear and switching. The latter finds favour particularly for its high efficiency and is discussed at the end of this chapter. Meanwhile, linear regulators will be covered in some detail. As might be expected, a linear voltage regulator circuit is one in which the principal devices are conducting at all times. Voltage regulators are either shunt or series with the main functional difference being that in a shunt regulator the controlling device is connected across (in shunt with) the load, while a series regulator is connected in series with the load (see Fig. 9.11).

> Exclude protection circuitry which is activated only under fault conditions.

Shunt voltage regulators

Simple breakdown diode regulator

> High currents require diodes with high power ratings.

A very simple shunt regulator in the form of a breakdown diode and current-limiting resistor has already been described in Chapter 1 and is redrawn in Fig. 9.12a. No feedback is involved and the performance is limited by the slope resistance, r_z, of the diode. Analysis of this regulator is simple as can be seen by referring to its equivalent circuit (Fig. 9.12b).

With the design values from Design Example 1.1 and Exercise 1.3 ($R_S = 330\ \Omega$ and $r_z = 40\ \Omega$), the RRF and r_0 can be calculated.

$$v_O = v_S - (i_O + i_Z)R_1$$

$$i_Z = \frac{v_O}{r_z}$$

$$\therefore v_O\left(1 + \frac{R_1}{r_z}\right) = v_S - i_O R_1$$

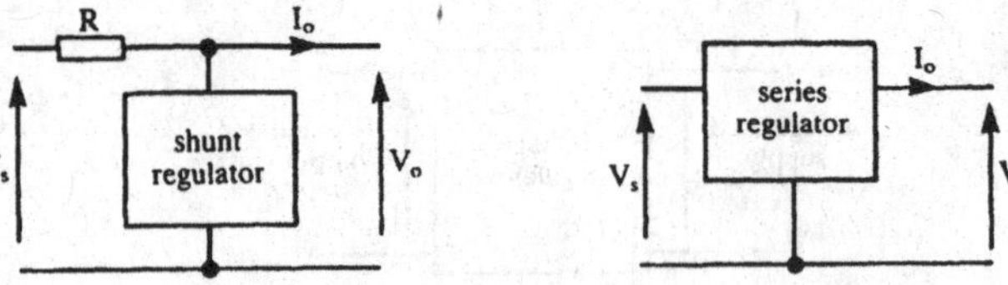

Fig. 9.11   Shunt and series voltage regulators.

186

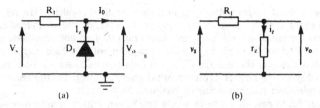

(a)   (b)

Fig. 9.12   (a) Circuit of simple regulator. (b) Equivalent circuit of (a).

$$v_O = \frac{r_z}{r_z + R_1} v_S - \frac{r_z R_1}{r_z + R_1} i_O$$

$$\text{RRF} = \frac{v_O}{v_S}\bigg|_{i_O = 0} = \frac{r_z}{r_z + R_1} = \frac{40}{40 + 330} = 0.108$$

and

$$r_O = -\frac{v_O}{i_O}\bigg|_{v_z = 0} = r_z \| R_1 = 40 \| 330 = 35.7\,\Omega$$

By incorporating a (power) transistor as shown in the rather more complex circuit of Fig. 9.13 the power dissipation is shifted from the breakdown diode to the transistor and, more important, a feedback mechanism is introduced which reduces r_O and enhances the RRF.

The feedback works as follows. If the output current I_O increases, the output voltage V_O falls due to the output resistance. Hence TR1's base-emitter voltage ($V_{BE1} = V_L - V_Z$) is caused to fall (assuming V_Z constant). The collector current of TR1 reduces, diverting input current from TR1 to the output and partially compensating for the assumed fall in V_O.

While this circuit may be modelled quite simply (the breakdown diode by its slope resistance and TR1 by a resistance r_{be} and current generator βi_b), the analytic expressions for r_O and RRF are complex and do not aid design of the circuit which, however, is straightforward. The regulated output voltage is simply the sum of V_Z and V_{BE}, so the diode is selected to have a breakdown voltage equal to 0.7 V less

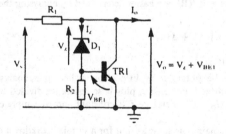

Fig. 9.13   BJT shunt regulator circuit.

than the required output voltage. As in the simple diode regulator, the resistor R_1 limits the total current drawn from the source. In providing maximum load current, a residual current of several milliamps should flow through both the diode and the transistor. Resistor R_2 (with V_{BE1} across it) together with the base current of TR1 establish the current through the diode. R_2 should be calculated so that the diode operates beyond the knee of its breakdown characteristic when the transistor current is at its minimum (corresponding to maximum load current).

(It is useful to note that a diode with a breakdown voltage of approximately 6 V exhibits a voltage temperature coefficient of approximately $+2\,\text{mV/}^\circ\text{C}$ which cancels the $-2\,\text{mV/}^\circ\text{C}$ temperature coefficient of V_{BE1} and yields an output voltage which is insensitive to temperature variations. However, the temperature coefficient of V_Z is dependent not only on V_Z but on the diode current as well. Consequently, for this cancellation to occur, I_Z should be well defined and constant.)

If the minimum load current is zero, TR1 must be rated such that it can pass a collector current at least equal to the maximum source current (which is greater than the maximum load current) and dissipate a power given by the product of output voltage and maximum source current. This poor efficiency for load currents below maximum is the major drawback of a shunt regulator. The maximum current through the diode is the sum of the maximum base current of TR1 and the current through R_2. D_1 must be selected accordingly. When resistor values have been calculated and devices selected, the circuit may be computer simulated to predict its performance.

(The output voltage of a shunt regulator may be made adjustable above and below V_Z by using rather more complex circuitry which is not treated here.)

A shunt regulator is ideal when stabilizing a small range of load current and it has the significant advantage that, under fault conditions with a short-circuit load, the transistor is not stressed; the load current is limited by R_1.

Regulator efficiency is defined as

$$\frac{\text{power in load}}{\text{power in load} + \text{dissipation in regulator}}$$

With the output short-circuit, the collector current of TR1 is zero.

Series voltage regulators

Emitter-follower regulator
Series voltage regulators are rather more popular than shunt regulators since, at output currents significantly less than the maximum output current, the dissipation is correspondingly lower. With internal feedback and amplification, they can provide good regulation and ripple reduction. However, the simplest series voltage regulator does not incorporate feedback and is simply a breakdown diode circuit buffered by an emitter-follower, TR1 in Fig. 9.14a. The output voltage is determined by the breakdown voltage of the diode ($V_O = V_z - V_{BE}$).

TR1 is referred to as the **series** or **pass** transistor.

Output resistance and RRF are easily determined by analysing the equivalent circuit of Fig. 9.14b.

$$r_o = r_e + (R_1 \| r_z)/(1 + \beta)$$

$$\text{RRF} = r_z/(R_1 + r_z)$$

See Exercise 4.3.

(These results can be obtained by inspection rather than analysis. The output resistance of an emitter follower is r_e plus base resistance divided by $(1 + \beta)$. RRF is calculated with $\Delta I_O = 0$. Therefore r_e is constant and we simply consider the R_1, r_z potential divider.)

While the RRF remains the same as that for a simple breakdown diode reference, r_O is nominally reduced by the β of the transistor.

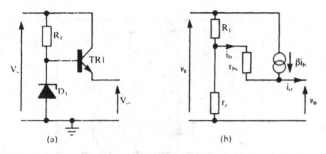

Fig. 9.14 Simple series voltage regulator (emitter-follower): (a) circuit diagram, (b) equivalent circuit.

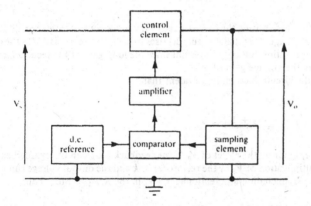

Fig. 9.15 Block diagram of feedback series voltage regulator.

Feedback series regulators using BJTs

A feedback series regulator (shown in block diagram form in Fig. 9.15) provides improved performance as well as the facility for adjusting the output voltage. A fraction (sample) of the output voltage is compared with a d.c. reference voltage. Any difference (error) is amplified and used to control the output voltage via the series element.

A single BJT may be used to provide both the differencing and the amplification (TR2 in Fig. 9.16). Sampling of the output voltage is performed by two resistors R_1 and R_2 acting as a potential divider with a ratio $k = R_2/(R_1 + R_2)$. A breakdown diode D_1 provides the d.c. reference voltage, V_z.

The action of the negative feedback loop is as follows. If the output voltage was to change due, say, to a change in load current, the feedback operates to partially compensate for the change. Assume that the load current is increased. Due to output resistance, the output voltage will fall, as will the voltage at the base of TR2. If the reference voltage remains constant, V_{BE} of TR2 reduces and the collector current of

Precision bandgap references are widely used, particularly in IC regulators. It is beyond the scope of this text to cover references, other than the breakdown diode.

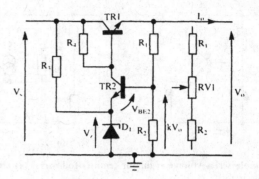

Fig. 9.16   Feedback series regulator using BJTs.

TR2 falls, in turn raising the base voltage of the series control transistor TR1. Hence its emitter voltage (the regulator output) rises to compensate partially for the original fall. Compensation can only be partial since the loop gain of the feedback circuit is limited by the voltage gain of TR2.

From the circuit diagram it is evident that

$$kV_O - V_{BE2} = V_Z$$

$$\therefore V_O = \frac{V_Z + V_{BE2}}{k}$$

Therefore, ignoring the effect of V_{BE2}, the feedback loop can be regarded as a step-up or multiplication of V_Z by the reciprocal of k and the output voltage can be made adjustable by including a potentiometer RV_1 in the $R_1 R_2$ resistor chain.

Small-signal analysis of series regulator
Assume that the breakdown diode is biased by TR2 emitter current only ($R_3 = \infty$) and is modelled by its slope resistance r_z; also that TR1 and TR2 are each modelled by r_{be} and current βi_b. The equivalent circuit of the regulator is given in Fig. 9.17.

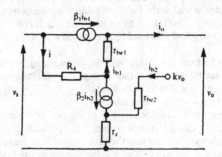

Fig. 9.17   Equivalent circuit of the feedback series regulator.

$$i_O = (1 + \beta_1) i_{b1}$$

$$i = i_{b1} + \beta_2 i_{b2}$$

$$v_S - i R_4 - i_{b1} r_{be1} = v_O$$

$$r_{be2} i_{b2} = k v_O - i_{b2}(1 + \beta_2) r_Z$$

$$\therefore i_{b2} = \frac{k v_O}{r_{be2} + (1 + \beta_2) r_Z}$$

$$\therefore v_S - i_{b1} R_4 - \beta_2 i_{b2} R_4 - i_{b1} r_{be1} = v_O$$

$$v_S - \frac{i_O}{(1 + \beta_1)}(R_4 + r_{be1}) - \beta_2 R_4 \frac{k v_O}{r_{be2} + (1 + \beta_2) r_Z} = v_O$$

$$\therefore v_O \left[1 + \frac{k \beta_2 R_4}{(1 + \beta_2)(r_{e2} + r_Z)} \right] = v_S - i_O \left[\frac{R_4}{(1 + \beta_1)} + r_{e1} \right]$$

Hence

$$RRF = \frac{v_O}{v_S} \bigg|_{i_O = 0} = \left[1 + \frac{k \beta_2 R_4}{(1 + \beta_2)(r_{e2} + r_Z)} \right]^{-1}$$

$$= \frac{r_{e2} + r_Z}{k R_4} \quad \text{if } \beta_2 \gg 1$$

and

$$r_O = -\frac{v_O}{i_O} \bigg|_{v_S = 0} = \frac{\dfrac{R_4}{(1 + \beta_1)} + r_{e1}}{\dfrac{k R_4}{(r_{e2} + r_Z)}} \quad \text{if } \beta_2 \gg 1 \quad \text{and} \quad k R_4 \gg (r_{e2} + r_Z)$$

$$\approx \frac{r_Z + r_{e2}}{k \beta_1}$$

These approximate results are sensible in that they show how the performance depends on the reference slope resistance, the amplifier gain (R_4/r_{e2}) and k. To maximize the loop gain, k should not be too small a fraction or, in other words, V_Z should not be chosen to be very much smaller than V_O. In fact k can be made equal to unity as far as the a.c. ripple is concerned by connecting a bypass capacitor across R_1.

The accuracy of small-signal analysis is questionable since regulators are usually required to operate over a range of current. Computer simulation should be used to check designs.

Power, current and voltage ratings of TR1 are determined by the maximum load current and the maximum voltage between collector and emitter, i.e. $V_S - V_O$, which must be greater than V_{BE} to correctly bias the transistor. However, under fault conditions such as a short-circuit load ($V_O = 0$ V), TR1 must be able to withstand the full input voltage. It is possible to parallel two or more series transistors to share the current and power dissipation requirements. In that case, a low-valued (typically less than 1 Ω) resistor is connected in series with each emitter to

The minimum voltage across the regulator, when it can still function, is called the **dropout voltage**.

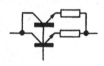

191

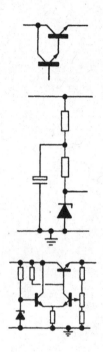

ensure that a transistor with a low V_{BE} does not carry significantly more than its share of load current. When the output current I_O is at its maximum, the base current of TR1 is at its maximum, $I_{O(max)}/\beta$. The collector current of TR2 when I_O is zero must be slightly greater than this otherwise, at full output current, TR2 would be cut off and control of the feedback loop would be broken. To ensure that the power dissipations in TR2 and D_1 are not excessive, two transistors in Darlington configuration will reduce the control (base) current requirement by a factor of β.

It is worth noting here the higher efficiency of a series regulator compared with that of a shunt regulator, particularly when the output current is less than the maximum. In a series regulator, as the load current falls, both load power and regulator dissipation decrease and the efficiency is approximately constant. However, in a shunt regulator, the load power decreases as load current falls but the regulator dissipation rises and very low efficiency can result.

The breakdown diode current has been shown derived from the input side of the regulator but V_S is contaminated by ripple. To substantially reduce the ripple appearing across the diode and hence at the output, filtering of the diode current may be used but it is simpler to provide the current from the regulated, relatively ripple-free output itself (Fig. 9.18). However, this connection relies on leakage current through TR1 to make the diode conduct when V_S is first applied (a start-up problem). This is bad design practice and the dependence on leakage is relieved by connecting a resistor R (a value in the tens of kilohms is satisfactory) between collector and emitter of TR1 to provide start-up current for the diode. As in the case of the shunt regulator, the diode breakdown voltage and current ($I_Z = I_{R3} + I_{E2}$) should be selected to cancel the temperature coefficient of V_{BE2} (approximately $-2\,mV/°C$) in order to minimize the temperature dependence of the output voltage. It is important to note that the base-emitter voltage of TR1 is within the feedback loop. If V_{BE1} changes due to temperature or current, the loop adjusts to compensate.

Rather than using a single, common-emitter amplifier TR2, a differential amplifier can provide high immunity to large temperature variations. In that case the d.c. reference should be biased for zero temperature coefficient.

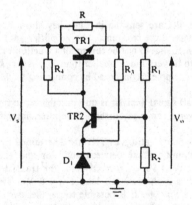

Fig. 9.18 Series regulator with the breakdown diode current derived from the output.

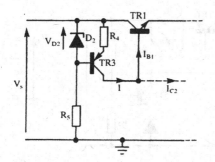

Fig. 9.19   Pre-regulator circuitry to reduce output ripple.

Despite biasing the breakdown diode from the output, the ripple reduction performance of the regulator is marred by the injection of ripple current from V_S via R_4 into the base of TR1. In order to reduce this, R_4 may be replaced by a constant current source (or pre-regulator), TR3 in Fig. 9.19. This current, I, is determined by the breakdown voltage of D_2 and the emitter resistor R_4 as

$$I = I_{C3} \approx I_{E3} = \frac{V_{D2} - V_{BE3}}{R_4}$$

D_2 is biased to give V_{D2} a temperature coefficient equal to that of V_{BE3} in order to keep constant the voltage across R_4. As well as reducing output ripple induced by the mechanism mentioned, the pre-regulator with its high output resistance has the further advantage of increasing the voltage gain of the amplifier which improves the RRF and r_O. There is a disadvantage in that the minimum voltage between regulator input and output ($V_S - V_O$) is increased by the voltage V_{D2}.

Overload protection

If the regulator output is subjected to a heavy load, e.g. a short-circuit, device ratings may be exceeded, resulting in catastrophic failure. Some means of overload protection therefore is usually incorporated in the regulator. There are two types of protection: current limiting and foldback protection. Several examples are considered.

Current limiting

The simplest form of current limiting is to incorporate a fuse in series with the unregulated supply. However, a fuse can take a considerable time to 'blow' by which time the damage is done. Alternatively, the fuse may be replaced by a resistor; dissipation in the series transistor(s) is reduced under short-circuit conditions but the performance of the power supply is adversely affected.

Figure 9.20a shows a more successful method of protection involving the additional components TR4 and R_6. TR4 does not conduct until its base-emitter voltage reaches approximately 0.7 V. Now $V_{BE4} = I_O R_6$ so when I_O reaches a level of $0.7/R_6$, TR4 conducts and diverts the pre-regulator current away from TR2

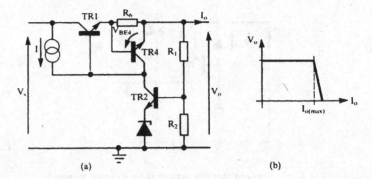

(a)   (b)

Fig. 9.20 (a) Current limit circuitry. (b) Output characteristic with current limiting.

which now fails to control the output voltage. The output current is limited to $I_{O(max)} = 0.7/R_6$ even when a short-circuit is applied at the output. The current-limited output characteristic is shown in Fig. 9.20b. Under short-circuit output conditions, the dissipation in the series transistor is the product of V_S and $I_{O(max)}$. The current limit can be made adjustable by including a potentiometer across R_6 which is now chosen to set the minimum current limit.

Foldback current limiting

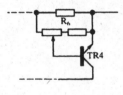

It is advantageous to create a 'foldback' output characteristic which gives a short-circuit current, I_{SC}, significantly less than $I_{O(max)}$. Dissipation in the series transistor is reduced. Figure 9.21b illustrates such a characteristic which can be achieved by adding two resistors, R_7 and R_8, to the previous current limit circuit as shown in Fig. 9.21a. Operation of the foldback protection circuit is as follows. When I_O is zero, V_{R6} is also zero and the base-emitter of TR4 is reverse biased by $V_{R7} = k'V_C$

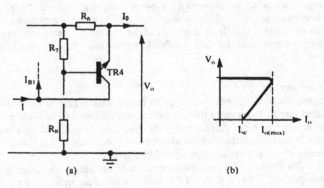

(a)   (b)

Fig. 9.21 (a) Foldback protection circuitry. (b) Output characteristic with foldback current limiting.

where $k' = R_7/(R_7 + R_8)$. As I_O increases from zero, the reverse bias reduces until TR4 conducts and limits I_O at $I_{O(max)}$, at which point

$$I_{O(max)}R_6 - k'(V_O + I_{O(max)}) = 0.7\,\text{V}$$

This gives

$$I_{O(max)} = \frac{0.7 + k'V_O}{(1 - k')R_6}$$

If an attempt is made to increase I_O beyond this level by, say, reducing the value of an external load resistor, the output voltage falls, as does the fraction $k'V_O$. Therefore the output current now required to keep TR4 conducting is reduced. In the limit when an output short-circuit ($V_O = 0$) is applied, the short-circuit current, I_{SC}, is determined by putting $V_O = 0$ in the previous expression.

$$I_{SC} = I_{O(max)}\Big|_{V_O = 0} = \frac{0.7}{(1 - k')R_6}$$

The ratio of maximum to short-circuit currents is given by

$$\frac{I_{O(max)}}{I_{SC}} = \frac{0.7 + k'V_O}{0.7}$$

which can be made quite significant. However, as this ratio is increased, k' and R_6 are also increased and the protection circuitry requires an increasingly significant voltage drop across it. In turn this reduces efficiency by demanding a higher input voltage to the regulator. The worked example below vividly illustrates the problem.

Design of a foldback protection circuit with V_O, $I_{O(max)}$ and I_{SC} specified would follow the procedure:

1. Calculate k' from the above equation rearranged as

$$k' = \frac{0.7\dfrac{I_{O(max)}}{I_{SC}} - 0.7}{V_O}$$

and select R_7 and R_8 accordingly.
2. Calculate R_6 from

$$I_{SC}\frac{0.7}{(1 - k')R_6}$$

A series regulator provides an output voltage of 12 V and a maximum output current of 1 A. Exercise 9.4

(a) Calculate the value of the current limiting resistor (R_6).
(b) What is the voltage dropped across R_6 when the output current is at its 1 A limit?
(c) Foldback protection circuitry is now added to reduce the short-circuit current to 0.5 A. By calculating k' and R_6 determine the voltage across R_6 at maximum current.
(d) Repeat (c) for a short-circuit current of 0.1 A.

Overvoltage protection

Some circuitry, 5 V TTL for example, may be destroyed if its power supply voltage is too high; this might happen if a fault were to develop in the regulator. A 'crowbar' circuit (Fig. 9.22) incorporating a thyristor or silicon-controlled rectifier (SCR) can be used to provide protection.

Refer to Bradley (1987), Chapter 1.

If the regulator voltage rises more than 0.7 V above the diode breakdown voltage, the SCR turns on rapidly and holds the output voltage near zero. The crowbar circuit can be reset only by turning off the supply.

Since the SCR can take a high continuous current and since the current limit in the regulator could become faulty, it is prudent to include a fuse between the unregulated supply and the regulator.

Series regulators using op-amps

An operational amplifier (op-amp) may be used to provide both the differencing and amplification as in Fig. 9.23. Since a high-gain amplifier requires only a very small differential input voltage, kV_0 is approximately equal to V_z and hence $V_0 = V_z/k$.

A potential problem with this circuit lies in the provision of d.c. supply for the op-amp. Since the output voltage of the op-amp is approximately 0.7 V more

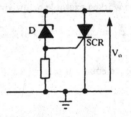

Fig. 9.22   Crowbar overvoltage protection circuit.

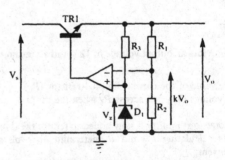

Fig. 9.23   Series regulator using an op-amp.

196

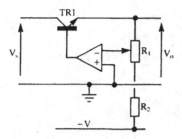

Fig. 9.24　Op-amp series regulator with output voltage adjustable down to 0 V.

positive than V_O, the regulated output voltage cannot be used as the op-amp's positive supply which therefore must be either the ripple-contaminated input voltage, V_s, or derived from it using a simple low-current regulator.

Another problem is that, without modification, the output voltage of this regulator cannot be adjusted to below the d.c. reference voltage; for $R_1 = 0$, $V_O = V_2$. In order to allow for adjustment down to zero voltage, it is necessary to provide a low-current, stabilized negative supply and use earth as the reference as shown in Fig. 9.24. This approach can also be used for a BJT version.

This feature is useful in bench power supplies.

Analysis of the op-amp series regulator of Fig. 9.24

Assume, with the breakdown diode current derived from the substantially constant output voltage, that the d.c. reference voltage is constant. The series transistor cannot be modelled only by r_{be} and βi_b; its output resistance $r(=1/h_{oe})$ should be included otherwise the RRF would be zero (see result). The op-amp is represented as an ideal voltage amplifier with voltage gain A. The equivalent circuit used in the analysis is shown in Fig. 9.25.

$$v_o + ir = v_s$$

$$i_b = -\frac{v_o(Ak + 1)}{r_{be}}$$

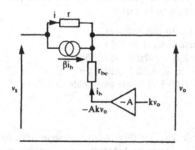

Fig. 9.25　Equivalent circuit of the op-amp regulator of Fig. 9.24.

$$i = i_o - (1 + \beta)i_b$$

$$\therefore \quad v_o = v_s - i_o r + (1 + \beta)i_b$$

$$= v_s - i_o r - \frac{(1 + \beta)(Ak + 1)r}{r_{be}} v_o$$

If $Ak \gg 1$ and $\dfrac{Akr}{r_e} \gg 1$

$$v_o = \frac{1}{Ak\dfrac{r}{r_e}} v_s - \frac{r_e}{Ak} i_o$$

r is the hybrid-π parameter
$r_{ce} = \mu r_e$ where $10^3 \leqslant \mu \leqslant 10^5$
(see Chapter 3).

$$\therefore \quad \text{RRF} = \frac{1}{Ak\dfrac{r}{r_e}} = \frac{1}{Ak\mu}$$

and $\quad r_o = \dfrac{r_e}{Ak}$

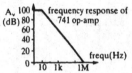

This analysis shows that, as might be expected, the performance of the regulator depends directly on the loop gain, Ak, of the feedback loop. While the open-loop voltage gain of an op-amp is very high at d.c., it falls off above typically 10 Hz. Although it is still very high at 50 or 100 Hz (ripple frequency), it is likely to be as low as 10 at 100 kHz. Therefore the ability of the regulator to attenuate wide-band noise, interference and the high-frequency components of the ripple is severely diminished. Also, the output resistance increases with frequency and variations in load current taken by one circuit cause power supply voltage changes which affect other circuits connected to the same supply. Careful decoupling of the regulator output and individual circuits overcomes problems at higher frequencies.

Regulator design examples

Perhaps the best way to summarize an area of study is to do an example. Here we will design two series voltage regulators: one using BJTs, the other an op-amp. In each case the specification to be met is

$V_O = 12$ V (adjustable to this voltage)
$I_O = 0$ to 1 A (current limited to 1 A)
Unregulated supply has regulation given by
$V_s = 25$ V with $I_o = 0$, $V_s = 20$ V with $I_o = 1$ A
and a peak-to-peak (100 Hz) ripple amplitude of 1 V.

BJT implementation

Referring to the circuit diagram (Fig. 9.26), the series transistor TR1 is specified first. It must be able to carry the 1 A full output current, withstand a collector-emitter voltage of at least 20 V and have a collector dissipation rating of 20 W when

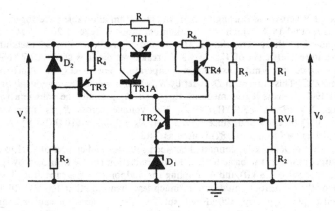

Fig. 9.26 A 12 V, 1 A series regulator using BJTs.

mounted on a suitable heatsink. The TIP31A, which has a minimum β of 25, is satisfactory. At an output current of 1 A, the maximum base current of TR1 is 1/25 A, or 40 mA. Since this level of current would cause a high dissipation in the amplifier transistor TR2, an additional transistor (TR1A) is added to TR1 to create a Darlington pair. Using a BC182 (50 V, 100 mA) with a minimum β of 80 reduces the maximum current required to control the series transistors to 40/80 mA, or 0.5 mA. The pre-regulator current should be greater than this to ensure that TR2 conducts at maximum output current. However, the choice of pre-regulator current, I_{C3}, is a compromise. If I_{C3} is low, then I_{C2} is low and hence r_{e2} is low and the r_0 and RRF are adversely affected; if I_{C3} is high, the dissipation in TR2 is high. A suitable compromise here would be to make I_{C3} equal to 5 mA.

Another BC182 may be used as the current limit transistor with the maximum current set to approximately 1 A by choosing $R_6 = 0.7/I_{O(max)} = 0.68\ \Omega$ (NPV).

The d.c. reference, breakdown diode D_1, should be selected and its bias current determined to give a temperature coefficient of around +2 mV/°C, cancelling that of V_{BE2}. A BZY88C6V2 diode (6.2 V) biased at 10 mA exhibits the correct temperature coefficient. Since $I_Z = 10$ mA $= I_{E2} + I_{R3}$ and I_{E2} will vary between 4.5 and 5 mA, a nominal I_{R3} of 5 mA should be provided by R_3 connected between the diode and the +12 V output. $R_3 = (12 - 6.2)/5 = 1.16$ kΩ. A 1.2 kΩ resistor (NPV) is used.

R_1 and R_2 are calculated next. The voltage at the base of TR2 is nominally 6.9 V (6.2 + 0.7) and the base current is less than 5/80 mA or 62.5 μA. The current through R_1 and R_2 should be greater than ten times this; allow approximately 1 mA. This is satisfied, and V_{B2} is approximately 6.9 V, if $R_1 = 5.1$ kΩ and $R_2 = 6.8$ kΩ. However, because of tolerances on resistor values and junction voltages, a potentiometer RV_1 should be introduced between R_1 and R_2 to allow the output voltage to be adjusted. If RV_1 is 470 Ω the output voltage should be adjustable by approximately ± 1 V.

Turning to the design of the pre-regulator, the voltage available for it should be considered. With an output voltage of 12 V, 0.7 V dropped across R_6 and allowing at least 2 V across the Darlington pair, the minimum allowable collector voltage of TR3 is around 15 V, which with a minimum input voltage of 20 V, leaves 5 V to accommodate the pre-regulator. This is not a problem since in matching the temperature coefficient of V_{BE3}, a 3.3 V breakdown diode is used. A BZY88C3V3, when biased at 5 mA, exhibits a temperature coefficient of approximately -2 mV/°C. This current in D_2 is set by R_5 which is calculated by considering that it has 19.2 V across it (the mean of $V_{s(min)}$ and $V_{s(max)}$ minus the voltage across D). Hence $R_5 = 19.2/5 = 3.9$ kΩ (NPV). The voltage across R_4 is approximately $(3.3 - 0.7)$ or 2.6 V and, since I_{E3} is 5 mA, $R_4 = 2.6/5 = 510$ Ω (NPV). A BC212 (complementary to the BC182) is used as TR3.

The resistor R (10 kΩ), connected between collector and emitter of TR1 to ensure start-up, completes the basic design. Ripple reduction can be enhanced by splitting R_5 (say 1.8 kΩ + 1.8 kΩ) and decoupling the midpoint with a capacitor. The reactance of the capacitor should be very much less than 1.8 kΩ at 100 Hz, the fundamental ripple frequency; 100 μF will suffice. Also, a capacitor can be connected between the output and the wiper of RV_1 for the same reason; again 100 μF would be satisfactory. Finally, the output is decoupled with another capacitor (100 μF).

Op-amp implementation

TR2 of the previous design is replaced by a 741 op-amp, the supply for which is taken from the unregulated input and earth (Fig. 9.27).

The Darlington pair, TR1 and TR1A, are retained as are the current limiter (TR4 and $R_6 = 0.68$ Ω) and the 10 kΩ start-up resistor, R. A 741 op-amp can easily supply the 0.5 mA maximum base current required by the base of TR1A.

D_1 is replaced by a BZY88C5V1 diode (5.1 V) which has a zero temperature coefficient at a current of 10 mA. This is derived by R_3, calculated as $(12 - 5.1)/10$ kΩ, or 680 Ω (NPV).

R_1 and R_2 may be increased since the input bias current (1.5 μA maximum for a 741) is much less than the base current of TR2. $R_1 = 68$ kΩ, $R_2 = 47$ kΩ and $RV_1 = 10$ kΩ are satisfactory.

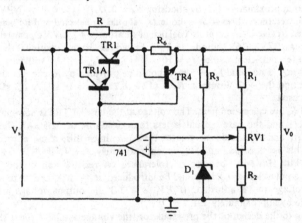

Fig. 9.27 Op-amp regulator circuit.

As in the previous design, the output is decoupled and the RRF enhanced by connecting a capacitor between the output and wiper of RV_1.

Both regulators have been constructed and tested.

Introduction to switching regulators

So far in this chapter only linear voltage regulators have been covered; another very important class of regulator – the switching regulator – must be introduced. The major advantage of a switching regulator (indeed, of all switching circuits) is high efficiency and this is achieved by switching the series transistor between ON (saturated) and OFF states, both of which are low dissipation. A basic switching regulator circuit is shown in Fig. 9.28.

The ON/OFF times of TR1 are controlled by a pulse-width-modulated (PWM) oscillator which has an output duty cycle (or mark-to-space ratio) determined by comparing a sample $(k = R_2/(R_1 + R_2))$ of the output voltage V_0 with a d.c. reference voltage. Operation of the circuit is best described by proceeding step-by-step through a cycle. At the start of the cycle TR1 is driven ON and the LC-filtered output voltage rises, as does kV_0. When kV_0 passes through a threshold voltage related to the reference, TR1 is switched OFF for the remainder of the oscillator cycle. However, the load current is not interrupted. The inductor L stores energy so current continues to flow while its magnetic field collapses. The diode completes the current path. By the start of the next oscillator cycle, kV_0 is below the threshold and TR1 is turned ON again. If the load current is increased it takes longer for V_0 to rise to the threshold and TR1 conducts for a longer fraction of the cycle; if the load current is reduced, the duty cycle is reduced. Hence the output voltage is maintained almost constant with only a small ripple amplitude provided that the LC (choke-input) filter components are selected appropriately.

D is reverse biased when TR1 is ON.

As well as the high efficiency (which can be greater than 90%) resulting from switching between low-dissipation states, a switching regulator has another significant advantage. Since the switching frequency can be high (over 100 kHz is possible), the values of inductance and capacitance are significantly smaller than those required for 50 Hz/100 Hz smoothing and the physical size of these components is similarly reduced. This may appear to be inconsequential since the unregulated

Switching frequencies are invariably above 20 kHz to avoid audible 'hum'.

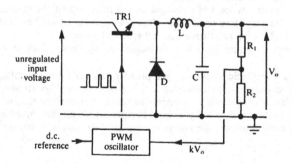

Fig. 9.28   Basic switching regulator circuit.

201

supply to which the regulator is connected would be expected to contain a transformer and filter components – the regulator L and C are additional.

However, the unregulated d.c. supply is not necessary and the input to a switching or switch-mode power supply (SMPS) may be taken directly from the mains itself, dispensing with the bulky mains transformer. In a SMPS the a.c. mains voltage is rectified, capacitor filtered and chopped at high frequency. A (high-frequency) transformer isolates input from output which, in its turn, is rectified and filtered. As in the switching regulator, the output voltage controls the duty cycle of the chopping transistor.

A disadvantage of switching techniques in power supplies is that they generate electromagnetic interference (EMI) due to high switching currents and fast switching. This EMI can disturb neighbouring equipment unless the SMPS input is adequately filtered and the unit screened.

It is beyond the scope of this text to develop further this discussion of switching regulators and power supplies; the interested reader must refer to more advanced publications.

Summary

The power supply must be the most ubiquitous electronic circuit since every piece of electronic equipment has to be supplied with power, usually d.c. at a controlled voltage. In this chapter we have covered in some detail design aspects of typical power supplies: the unregulated supply with its transformer, rectifier and filter, as well as the regulator circuit whose function is to stabilize the output voltage to a high degree of precision.

Half-wave, full-wave and bridge rectifier configurations have been described together with capacitor and choke-input filters at a level of detail which permits calculation of component values as well as their voltage and current ratings.

Voltage regulator circuit design has been concentrated on series regulators incorporating negative feedback, although shunt regulators and the use of an emitter-follower to reduce output resistance are covered briefly. Pre-regulation and protection circuits have been introduced.

The value of small-signal a.c. analysis of regulator circuits to predict performance is rather limited. It may serve only to confirm that performance is degraded by reference slope resistance and enhanced by the loop gain of the feedback. Since power supplies may have to accommodate a wide range of load current, design, simulation and testing must cover the full specified range.

Two design examples have been presented using bipolar transistors and an op-amp.

In the space available, it has not been possible to cover aspects of integrated circuit regulators from precision bandgap d.c. references to regulator circuits requiring only a minimum of external components to realize high-performance d.c. voltage supplies. However, many of the discrete component circuit techniques introduced in this and earlier chapters are just as relevant to their integrated counterparts.

Problems

9.1 A d.c. power supply has a no-load output voltage of 24 V and a full-load voltage of 22 V. What is the percentage regulation?

9.2 A capacitor filter is used to smooth the output of a bridge rectifier. The peak output voltage is 22 V at a mean load current of 100 mA and the peak-to-peak ripple voltage is to be less than 1.5 V. Assuming a mains frequency of 50 Hz, calculate the value of capacitor required.

9.3 What value of capacitor would be needed to meet the specification given in Problem 9.2 if a half-wave rectifier is used?

9.4 An unregulated d.c. voltage supply is required to supply an output current ranging from 50 mA to 200 mA at an output voltage (assumed constant for currents above the critical level) of 12 V. A choke-input filter is used to reduce the output ripple to 1 V peak-to-peak. Calculate the values of inductor and capacitor.

9.5 Using a breakdown diode and resistor, design a simple voltage regulator to provide an output of approximately 5 V at a maximum current of 20 mA from a 14 V source (assumed constant voltage). Ensure that a minimum current of 5 mA flows through the diode.

9.6 In the circuit of Problem 9.5, the 14 V source is now assumed to be contaminated with 1 V peak-to-peak ripple. If the diode slope resistance is 30 Ω, calculate the output ripple. What is the output resistance?

9.7 Design a BJT shunt regulator to provide an output voltage of approximately 7 V at an output current ranging between 100 mA and 150 mA. The input voltage is 14 V.

9.8 Design a simple emitter-follower series regulator to provide approximately 4 V at an output current ranging from zero to 100 mA. The input voltage is 10 V.

9.9 For the circuit designed in Problem 9.8, estimate the r_o and RRF. Assume a load current of 50 mA, a diode slope resistance of 30 Ω and a transistor β of 50.

Appendix A
Preferred Values for Passive Components

It is impossible for any laboratory, large or small, to keep in stock every possible value of resistor between, for example, 1 Ω and 10 MΩ. Not only would the cost of each resistor be great but the storage space required would be immense!

Low-cost, as distinct from precision, resistors are manufactured within certain tolerance limits, for example ±10%, ±5% or ±2%, and it is unnecessary to produce resistors whose tolerance bands overlap significantly. We therefore have ranges of preferred values for resistors and can select the nearest preferred value (NPV) to that arising from a design calculation.

Consider the arithmetic series of nominal resistor values 1, 2, 3, ..., 8, 9 each with a tolerance of ±10%.

nominal value	1	2	3	...	7	8	9
maximum value	1.1	2.2	3.3	...	7.7	8.8	9.9
minimum value	0.9	1.8	2.7	...	6.3	7.2	8.1

It is apparent that, at the low end of the range, there are significant gaps between the maximum value of one component and the minimum value of the one above. At the high end of the range the tolerance limits overlap.

To give minimum gaps or overlaps between adjacent preferred values when tolerance is taken into account, preferred value ranges are based on a geometric series rather than an arithmetic one.

The values forming the geometric progression given in Table A.1 are related as powers of $10^{1/24}$, each rounded to two significant figures, i.e. $10^{1/24}$, $10^{2/24}$, $10^{3/24}$, ..., $10^{22/24}$, $10^{23/24}$. This is known as the E24 series which accommodates ±5% tolerance. Shown in bold type in Table A.1 is the E12 subset $10^{1/12}$, $10^{2/12}$, $10^{3/12}$, ..., $10^{10/12}$, $10^{11/12}$, which is appropriate to ±10% tolerance components.

There are 24 basic values in the E24 series, 12 in the E12 series, and so on.

Table A.1 E24 Series of preferred values (with the E12 series in bold type)

1.0	1.1	**1.2**	1.3	**1.5**	1.6
1.8	2.0	**2.2**	2.4	**2.7**	3.0
3.3	3.6	**3.9**	4.3	**4.7**	5.1
5.6	6.2	**6.8**	7.5	**8.2**	9.1

Exercise A.1 By considering the E12 series subject to ±10% tolerance, show that the gaps and overlaps are minimal compared with the arithmetic series of single decimal digits.

For closer tolerances such as ±2% and ±1%, E48 and E96 series ($10^{n/48}$ and $10^{n/96}$) are used.

So far we have considered only one decade range of resistance. Other decades are covered simply by multiplying the basic values by the appropriate integer power of ten, for example 3.3 Ω, 33 Ω, 330 Ω, 3.3 Ω, 33 Ω, etc.

Capacitors and inductors are also supplied with preferred values in a geometric progression often restricted to the E6 or even the E3 series (1.0, 2.2, 4.7) for certain types.

Appendix B
Transient response of R–C circuits

We wish to find the output voltage (V_o) as a function of time for an input step of voltage, amplitude V, applied at time $t = 0$ to the two simple resistor-capacitor circuits of Fig. B.1. The recognized method of analysis is by solving the differential equation for each circuit, either by the classical approach of considering the complementary function and particular integral or by using Laplace transforms, to yield the following results.

The classical solution technique is well covered in

Fidler and Ibbotson (1989), pp. 43–64, and

Kuo (1966), Chapters 4 and 5.

Fidler and Ibbotson (1989) also present the Laplace transform method in Chapter 4.

For the circuit of Fig. B.1a,
$$V_o = V \exp(-t/\tau) \tag{B.1}$$
and for Fig. B.1b,
$$V_o = V[1 - \exp(-t/\tau)] \tag{B.2}$$
In each case the time constant,
$$\tau = RC \tag{B.3}$$

These output responses are plotted in Fig. B.2. Putting numbers into Equations B.1

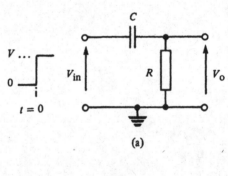

(a)

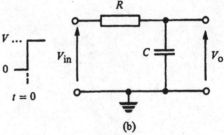

(b)

Fig. B.1

and B.2 yields several useful results which should be committed to memory; these are listed on p. 208.

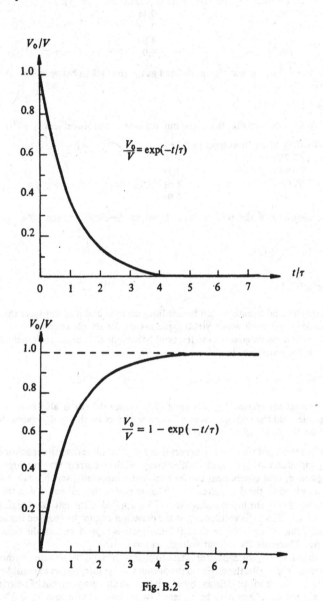

Fig. B.2

For the $C-R$ circuit of Fig. B.1a, the initial slope of the waveform is $-V/\tau$.

V_o falls to 36.8% of its initial level after one time constant (τ).

90%	0.1τ
10%	2.3τ
1%	4.6τ
0.1%	6.9τ

The fall time (t_f) of the waveform, defined as the time taken between 90% and 10% levels, is

$$t_f = (2.3 - 0.1)\tau = 2.2\tau \qquad \text{(B.4)}$$

For the $R-C$ circuit of Fig. B.1b, the initial slope of the waveform is $+V/\tau$.

V_o rises to 10% of its final level in 0.1τ.

63.2%	1.0τ
90%	2.3τ
99%	4.6τ
99.9%	6.9τ

The rise time (t_r) of the waveform, defined as the time between 10% and 90% levels, is

$$t_r = (2.3 - 0.1)\tau = 2.2\tau \qquad \text{(B.5)}$$

Intuitive solution

Solving differential equations can be a tedious exercise and it is fortunate that there is an intuitive approach which yields rapid results for simple circuits.

The equation which governs the terminal behaviour of a capacitor (equivalent to Ohm's law for a resistor) is

$$\frac{dV_c}{dt} = \frac{I_c}{C} \qquad \text{(B.6)}$$

In any practical system I_c is always finite. Hence dV_c/dt is also always finite, which implies that the voltage across a capacitor cannot be changed instantaneously.

Consider the $C-R$ circuit of Fig. B.1a.

1. Before the input voltage step occurs (i.e. $t < 0$), assume that all capacitor charging (or discharging) current has died away. With no current flowing through the capacitor, that component can be removed temporarily and V_o, for, $t < 0$, is determined by the d.c. voltage at the lower end of the resistor, 0 V in this case.
2. At time $t = 0$, the input voltage step (V) is applied to the left-hand plate of the capacitor. Since the voltage across the capacitor cannot be changed instantaneously, the voltage step on the left-hand plate is reproduced on the right-hand plate. Therefore the output voltage, at $t = 0$, is V.
3. At time $t = 0+$, the capacitor is free to charge through the resistor. Rather than attempt to describe the effect on the output voltage directly, let us consider what happens as t tends to infinity, by which time all charging current has died away. Again the capacitor may be replaced by an open-circuit and $V_o = 0$ V. This final level is called the **aiming potential**.

4. We now have three points on the time waveform for V_o. Between $t = 0$ and $t = \infty$ there can only be a smooth change of V_o; there cannot be an abrupt discontinuity in V_o since, after the input step at $t = 0$, the input voltage is held at a steady level. The only solution to this region is of the form $[\exp(-t/\tau)]$ and $V_o = V[\exp(-t/\tau)]$ since the output is obviously proportional to the amplitude of the input voltage step.

5. It remains to calculate the time constant (τ) as a function of the resistors and capacitors in the circuit. In this case there is only one resistor R and one capacitor C and τ is simply equal to RC.

This systematic application of the five steps (1) to (5) yields the analytic expression for the output voltage without recourse to solving a differential equation.

$$V_o = V[\exp(-t/\tau)] \qquad \tau = RC \qquad \text{(B.7)}$$

Although this method may appear lengthy when written down, it can yield rapid results after a little practice. It is left to the reader to apply this technique to the R–C circuit of Fig. B.1b and justify the result given in Equation B.2.

Determination of time constant

In the preceding example determining the time constant was simple since there were only two components. In more complex circuits with several resistors, how may the time constant be calculated?

Consider the circuit of Fig. B.3a. To determine the natural response of the circuit, replace the input voltage source with a short-circuit. (If the circuit was current-

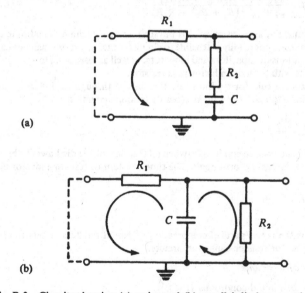

(a)

(b)

Fig. B.3 Circuits showing (a) series and (b) parallel discharge paths.

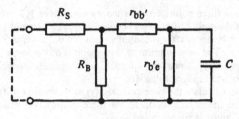

Fig. B.4

driven, the input current source should be open-circuited.) Also imagine that the capacitor has been charged up to an arbitrary potential. The capacitor now discharges through a single path, i.e. through resistors R_1 and R_2 in series. The time constant for this circuit is therefore $C(R_1 + R_2)$.

For the circuit of Fig. B.3b, the capacitor discharges through two parallel paths: via R_1 and R_2 in parallel. The time constant of this circuit is therefore $C(R_1 \| R_2)$.

In a more complex example such as the input section of the equivalent circuit of a BJT amplifier (Fig. B.4), the time constant can still be determined by inspection as

$$\tau = C[r_{b'e} \| (r_{bb'} + R_B \| R_S)] \tag{B.8}$$

This is a very powerful technique which rapidly gives results in a useful form. For example, it would be rather difficult to interpret the result of Equation B.8 if analysis had produced the expanded form

$$\tau = C\frac{r_{b'e}[(R_S + R_B)r_{bb'} + R_S R_B]}{(r_{b'e} + r_{bb'})(R_S + R_B) + R_S R_B} \tag{B.9}$$

Unfortunately, application of this intuitive analysis technique and time constant determination is not readily extensible to circuits containing more than one capacitor, or combinations of capacitors and inductors, as well as resistors. However it is useful for circuits with a single inductor plus resistors.

The intuitive rule for an inductor, the current through an inductor cannot be changed instantaneously, follows from its terminal equation

$$V_L = L\frac{dI_L}{dt} \tag{B.10}$$

and under steady-state conditions (when all transients have died away), the inductor may be considered a short-circuit. The time constant of resistor-inductor circuits is given by

$$\tau = \frac{L}{R} \tag{B.11}$$

where R is the series/parallel combination of circuit resistances determined in the same way as for resistor-capacitor circuits.

Summary of technique

1. Establish initial conditions ($t < 0$).
2. Determine result of input transient ($t = 0$).

3. Work out final steady-state or aiming potential $(t \rightarrow \infty)$.
4. Interpolate between (1) and (2) with $[\exp(-t/\tau)]$ or $[1 - \exp(-t/\tau)]$.
5. Calculate time constant (τ).

Problems

The circuits below have voltage inputs applied as shown. Draw the output waveform (V_{out}) labelling time constants and voltage levels.

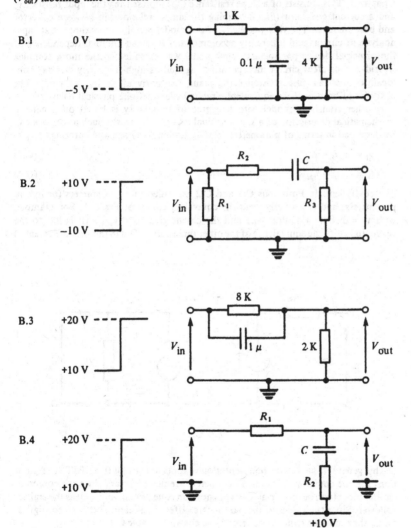

Appendix C
h-parameter modelling of BJTs

A simple small-signal a.c. model for a BJT in common-emitter was introduced in Chapter 3. This consists of a slope resistance (r_{be}) representing the input resistance, and a current generator (βi_b) describing the simple relationship between collector and base currents. The very simplicity of this model permits approximate but rapid analysis of circuits and inherently accommodates β spread and bias dependence of the principal device parameters. This model was extended to the more complex hybrid-π equivalent circuit incorporating, as well as high-frequency effects, non-idealities such as base resistance ($r_{bb'}$) and the Early effect ($r_{b'c}$ and r_{ce}). The elements of both of these equivalent circuits have a definite physical basis.

Another transistor model with a degree of popularity is based on a general mathematical description of a four-terminal two-port network. Such a network can be described in terms of parameters relating terminal voltage and currents;

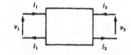

$$v_1 = h_{11}i_1 + h_{12}v_2 \tag{C.1}$$

$$i_2 = h_{21}i_1 + h_{22}v_2 \tag{C.2}$$

In all, there are six ways of relating the input and output variables (v_1, i_1, v_2, i_2). Kuo (1966, pp. 253–256) describes the four most useful sets of parameters, including the h-parameters.

The coefficients in Equations C.1 and C.2 are called the h-parameters (or hybrid parameters) and collectively yield the equivalent circuit of Fig. C.1. For example, with $v_2 = 0$, both Equation C.1 and the circuit give $h_{11} = v_1/i_1$. (It is left to the reader to justify the equivalence of the other three parameters to the circuit elements.)

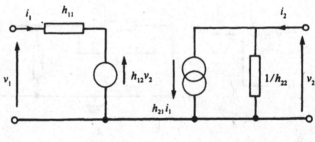

Fig. C.1

This general h-parameter representation is applicable to all three BJT configurations but the common-collector h-parameters are not necessary since the common-emitter model can be used in analysis of common collector circuits such as the emitter follower. Since the values of the parameters differ significantly between configurations, they must be denoted differently as shown in Table C.1.

Using tables, it is possible to convert from CE h-parameters to CB parameters, and vice versa.

Table C.1

Parameter	Dimensions	CB	CE
h_{11} input	resistance	h_{11} or h_{ib}	h'_{11} or h_{ie}
h_{12} reverse	–	h_{12} or h_{rb}	h'_{12} or h_{re}
h_{21} forward	–	h_{21} or h_{fb}	h'_{21} or h_{fe}
h_{22} output	conductance	h_{22} or h_{ob}	h'_{22} or h_{oe}

Note that the parameters have mixed dimensions – hence their name, **hybrid parameters**: h_i and h_o represent the input resistance and output conductance respectively, h_r the internal voltage feedback ratio from output to input, and h_f the forward current ratio. Clearly, h_{fb} and h_{fe} are equivalent to the current gains α and β.

It is important to note that the h-parameters vary with bias (I_C and V_{CE}) and temperature. Also they are specified as real numbers and therefore do not include high-frequency effects. They are valid only for low-frequency (audio) applications. If quoted at all in manufacturers' data, they are specified as having been measured at a certain bias, temperature and frequency.

Typical values of CE h-parameters for the BC 107B transistor at $I_C = 2\,\text{mA}$, $V_{CE} = 5\,\text{V}$ and $f = 1\,\text{kHz}$ are:

$$h_{ie} = 4.5\,\text{k}\Omega, h_{re} = 2 \times 10^{-4}, h_{fe} = 330, h_{oe} = 3 \times 10^{-5}\,\text{S}.$$

Since a specific transistor is most unlikely to exhibit the typical parameter values of its type, analysis using the full set of four h-parameters is of little value unless the parameters have actually been measured at the correct bias for the particular BJT under consideration. Therefore, a simplified model, using a dominant subset of the parameters, would be useful. It has been shown that the elements h_{oe} and h_{re} can be omitted without introducing more than 10% error if the collector load resistor (R_C) is lower in value than $1/h_{oe}$ by a factor of ten. The resulting model, with h_{ie} and h_{fe} remaining, is identical in form to the simple model of Chapter 3, h_{fe} is obviously equivalent to β, but does h_{ie} equal r_{be}? Let us calculate values using the BC 107B data given above.

At $I_C = 2\,\text{mA}$, $r_{be} = (1 + \beta)r_e = (1 + h_{fe})r_e = 331 \times 12.5 = 4.14\,\text{k}\Omega$

and

$$h_{ie} = 4.5\,\text{k}\Omega$$

The discrepancy of $360\,\Omega$ is due to $r_{bb'}$ which, although included in the full hybrid-π equivalent circuit, is omitted for the simple model.

Thus

$$h_{ie} = r_{bb'} + (1 + h_{fe})r_e \tag{C.3}$$

Using this relationship is a useful way of estimating $r_{bb'}$ from data if h_{ie}, h_{fe} and I_C are given, and is accurate if calculations are performed on measured data for a particular device.

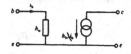

Table C.2

Amplifier parameter	Result using full h-parameter model	Result using simplified model	Refer to Equation
Common-emitter amplifier			
A_V	$\dfrac{-h_{fe}R_C}{h_{ie}(1 + h_{oe}R_C) - h_{fe}h_{re}R_C}$	$\dfrac{-h_{fe}R_C}{h_{ie}}$	3.7
r_{in}	$h_{ie} - \dfrac{h_{fe}h_{re}R_C}{1 + h_{oe}R_C}$	h_{ie}	3.10
g_{out}	$h_{oe} - \dfrac{h_{fe}h_{re}}{h_{ie}} + \dfrac{1}{R_C}$	$\dfrac{1}{R_C}$	3.12
Series feedback amplifier			
g_T	$\dfrac{h_{fe} - \dfrac{h_{oe}[h_{fe}R_C + (1 + h_{fe})R_E]}{1 + h_{oe}(R_C + R_E)}}{h_{ie} + (1 + h_{fe})R_E - \dfrac{h_{re} + h_{oe}R_E}{1 + h_{oe}(R_C + R_E)}[h_{fe}R_C + (1 + h_{fe})R_E]}$	$\dfrac{h_{fe}}{h_{ie} + (1 + h_{fe})R_E}$	4.6
r_{in}	$h_{ie} + (1 + h_{fe})R_E - \dfrac{h_{re} + h_{oe}R_E}{1 + h_{oe}(R_C + R_E)}[h_{fe}R_C + (1 + h_{fe})R_E]$	$h_{ie} + (1 + h_{fe})R_E$	4.4
$r_{out(t)}$	$\dfrac{(1 + h_{fe})R_E + h_{ie}(1 + h_{oe}R_E) - (1 + h_{fe})h_{re}R_E}{h_{oe}\left[h_{ie} + R_E - \dfrac{h_{re}h_{fe}}{h_{oe}}\right]}$	$\dfrac{1}{h_{oe}}\left[1 + \dfrac{h_{fe}R_E}{h_{ie} + R_E}\right] + h_{ie} /\!/ R_E$	4.10
Emitter follower			
A_V	$\dfrac{(1 + h_{fe})R_E}{h_{ie}(1 + h_{oe}R_E) + (1 - h_{re})(1 + h_{fe})R_E}$	$\dfrac{(1 + h_{fe})R_E}{h_{ie} + (1 + h_{fe})R_E}$	4.14
r_{in}	$h_{ie} + \dfrac{(1 - h_{re})(1 + h_{fe})R_E}{1 + h_{oe}R_E}$	$h_{ie} + (1 + h_{fe})R_E$	4.13
r_{out}	$\dfrac{h_{ie}R_E}{h_{ie}(1 + h_{oe}R_E) + (1 - h_{re})(1 + h_{fe})R_E}$	$\dfrac{h_{ie}}{1 + h_{fe}} /\!/ R_E$	4.15
Shunt feedback amplifier			
r_T	$-R_B\left[\dfrac{\dfrac{(1 + h_{fe})R}{h_{ie} + R_B} - \dfrac{R}{R_B}}{1 + \dfrac{(1 - h_{re})(1 + h_{fe})R}{h_{ie} + R_B}}\right]$ where $R = R_C /\!/ 1/h_{oe}$	$\dfrac{-R_B}{1 + \dfrac{h_{ie} + R_B}{(1 + h_{fe})R_C}}$ if $h_{ie} \ll h_{fe}R_B$	4.23
r_{in}	$\dfrac{[h_{ie} - h_{re}(1 + h_{fe})R][R_B + R(1 - h_{re})]}{(1 + h_{fe})(1 - h_{re})R + R_B + h_{ie}}$	$\dfrac{R_B + R_C}{1 + \dfrac{R_B + (1 + h_{fe})R_C}{h_{ie}}}$	4.24
r_{out}	$\dfrac{R_B + h_{ie}}{(1 + h_{fe})(1 - h_{re})} /\!/ \dfrac{1}{h_{oe}} /\!/ R_C$	$\dfrac{R_B + h_{ie}}{1 + h_{fe}} /\!/ R_C$	4.25

Circuit analysis using h-parameters

Analysis of equivalent circuits incorporating the full h-parameter model is much more tedious than with the simplified model. Presented in Table C.2 are the results of analysis of several amplifier configurations. First the full model is used, then h_{oe} and h_{re} are omitted yielding (as would be expected) results exactly corresponding to those using the $r_{be}/\beta i_b$ model for which cross-references to Chapters 3 and 4 are provided. It is left to the reader to verify these results.

Results with the full model are of little value for design purposes – they are far too complex.

The obvious complexity of the expressions involving the full set of h-parameters restricts their use to analysis where maximum accuracy is required and the parameters of the particular device have been measured. The full h-parameter model can have little application in the design process where initial concepts assuming ideal devices are tempered by including **first order** limitations. For this, h_{ie} and h_{fe} are adequate in most circumstances but it is important to remember that h_{fe} is subject to a wide spread and that h_{ie} is dependent on bias and h_{fe} (see Equation C.3). Hence typical parameter values extracted from data sheets have to be used with caution.

A knowledge of h_{oe} would be useful for current sources.

Although the h_{ie}/h_{fe} model is equivalent, the author prefers to use the $r_{be}/\beta i_b$ model. Since r_{be} as such is never quoted in data (it has to be calculated from β and bias), there is never the temptation to regard BJT input resistance as an invariant parameter. When the simpler $r_{be}/\beta i_b$ model is extended to the full hybrid-π equivalent circuit, the bias dependence of r_{ce} and $r_{b'c}$ ($r_{ce} = \mu r_e$ and $r_{b'c} = \mu \beta r_e$) is immediately apparent unlike h_{oe} and h_{re} for which graphical data have to be consulted.

References

Attikiouzel, J. and Jones, P.E. (1991) *C for Electrical and Electronic Engineers*, Prentice Hall.

Bannister, B.R. and Whitehead, D.G. (1991) *Transducers and Interfacing*, 2nd edn, Chapman and Hall, London.

Bissell, C.C. (1988) *Control Engineering*, Chapman and Hall, London.

Bradley, D.A. (1987) *Power Electronics*, Chapman and Hall, London.

Clayton, G.B. (1975) *Linear Integrated Circuit Applications*, Macmillan, London.

Clayton, G.B. (1979) *Operational Amplifiers*, 2nd edn, Butterworth, London.

Fidler, J.K. and Ibbotson, L. (1989) *Introductory Circuit Theory*, 2nd edn, McGraw-Hill, Maidenhead.

Goodge, M.E. (1990) *Analog Electronics*, Macmillan, London.

Horrocks, D.H. (1990) *Feedback Circuits and Operational Amps*, 2nd edn, Chapman and Hall, London.

Kuo, F.F. (1966) *Network Analysis and Synthesis*, Wiley.

Millman, J. and Halkias, C.C. (1972) *Integrated Electronics*, McGraw-Hill.

Millman, J. and Grabel, A. (1987) *Microelectronics*, 2nd edn, McGraw-Hill.

Morant, M.J. (1990) *Integrated Circuit Design and Technology*, Chapman and Hall, London.

O'Reilly, J.J. (1989) *Telecommunication Principles*, 2nd edn, Chapman and Hall, London.

Sangwine, S.J. (1987) *Electronic Components and Technology*, Chapman and Hall, London.

Schade, O.H. (1943) *Analysis of rectifier operation*, Proc. IRE, 31, 343–346.

Sparkes, J.J. (1987) *Semiconductor Devices*, Chapman and Hall, London.

Stonham, T.J. (1987) *Digital Logic Techniques*, 2nd edn, Chapman and Hall, London.

Texas Instruments (1989) *Linear Circuits Data Book 3*.

Answers to problems

1.1 (a) diode forward biased; $V_x = +6.2\,\text{V}$, $V_y = +6.9\,\text{V}$
    (b) diode reverse biased; $V_x = 0\,\text{V}$, $V_y = +10\,\text{V}$
1.2 $r_d = 12.5\,\Omega$
1.3 $\Delta V = -34.6\,\text{mV}$
1.4 $|BV| = 6.2\,\text{V}$, $R_s = 220\,\Omega$, $P = 248\,\text{mV} < 400\,\text{mW}$ rating
    Ripple reduction factor $= 0.031$ (or its reciprocal, 32.4)
1.5 $\beta = 124$
1.6 $\alpha = 0.9967$
1.7 For an n-p-n transistor, there is hole current injected from base to emitter, also
    recombination in the base region. Both serve to reduce the current transfer from
    emitter to collector.
1.8 See text
1.9 See text
1.10 Length $= 22.7\,\text{mil}$
1.11 See text

2.1 $r_{in} \geqslant 49\,\text{k}\Omega$
2.2 $r_{out} \leqslant 101\,\Omega$
2.3 $I_C = 3\,\text{mA}$ (approx.), $V_{CE} = +6\,\text{V}$ (approx.)
2.4 $I_C = 2.2\,\text{mA}$ (approx.), $V_{CE} = +6.7\,\text{V}$ (approx.)
2.5 $I_C = 4.3\,\text{mA}$ (approx.), $V_{CE} = +9.2\,\text{V}$ (approx.). [Remember V_{BE}!]
2.6 $I_C = 4.9\,\text{mA}$ (approx.), $V_{CE} = +6.3\,\text{V}$ (approx.)
2.7 $R_C = 8.2\,\text{k}\Omega$, $R_B = 1.2\,\text{M}\Omega$ (nearest preferred values, $\beta_{typ} = 200$)
2.8 (a) $V_C = 9\,\text{V}$, $V_B = 4\,\text{V}$, $V_E = 3.3\,\text{V}$ (all approximate since β not given)
    (b) V_E fixed at 3.3 V, $3.3\,\text{V} \leqslant V_C \leqslant 15\,\text{V}$ (fully saturated to OFF)
2.9 $V_E \gg \Delta V_{BE}$, 3 V o.k.; $I_C = 2\,\text{mA} \rightarrow I_{B(max)} = 40\,\mu\text{A}$, say;
    $I_{R2} \approx 300\,\mu\text{A}$, between 5 and 10 times $I_{B(max)}$, o.k.
2.10 $R_E = 2.2\,\text{k}\Omega$, $R_C = 6.8\,\text{k}\Omega$, $R_1 = 82\,\text{k}\Omega$, $R_2 = 22\,\text{k}\Omega$
2.11 $V_C = 6\,\text{V}$, $V_B = 2.7\,\text{V}$, $V_E = 2\,\text{V}$ (assuming β to be infinite)
2.12 R_{E1}
2.13 $A_V = 758$, $f_o = 3.1\,\text{Hz}$
2.14 $A_V = 90.9$, $f_o = 1.45\,\text{Hz}$

3.1 $A_V = -120$, $r_{in} = 831\,\Omega$, $r_{out} = 1\,\text{k}\Omega$
3.2 $A_V = -238$, $r_{in} = 1016\,\Omega$, $r_{out} = 2.7\,\text{k}\Omega$
3.3 $A_V = -160$, $r_{in} = 953\,\Omega$, $r_{out} = 1\,\text{k}\Omega$
3.4 $\beta = 122.2$, $r_e = 12.5\,\Omega$
    $v_o/v_s = -155.7$, $f_o = 75.1\,\text{Hz}$
3.5 $v_o/v_s = -148.7$, $f_o = 71.8\,\text{Hz}$
3.7 $A_V = +237.6$, $r_{in} = 11.5\,\Omega$, $r_{out} = 2.7\,\text{k}\Omega$
3.8 (a) $R_C = 2\,\text{k}\Omega$ (b) $R_E = 2\,\text{k}\Omega$
    (c) $A_V = -80$ (d) $r_{in} \approx 2.5\,\text{k}\Omega$

3.9 $r_e = 12.5\,\Omega$ and $g_m = 80\,\text{mS}$, $r_{b'e} = 1.2625\,\text{k}\Omega$, $r_{ce} \approx 125\,\text{k}\Omega$, $r_{b'c} \approx 12.5\,\text{M}\Omega$, $r_{bb'} = 100\,\Omega$, $C_{b'c} = 5\,\text{pF}$, $C_{b'e} = 101\,\text{pF}$.

3.10 For an n-p-n common-base amplifier (Fig. 3.5), if V_{in} rises, V_{BE} falls, I_C falls and V_{out} rises. Therefore, non-inversion.

3.11 $A_V = 1765$, $r_{in} = 25.25\,\text{k}\Omega$, $r_{out} = 4.7\,\text{k}\Omega$, $f_o = 3.2\,\text{Hz}$

4.1 $A_V = -1.785$, $r_{in} = 8.32\,\text{k}\Omega$

4.2 $A_V = -1.975$, $r_{in} = 3.42\,\text{k}\Omega$

4.3 $A_V = -0.45$, $r_{in} = 334\,\text{k}\Omega$, $r_{out} \approx 1.5\,\text{k}\Omega$ Ref. Fig. 4.6, $A_{V(d.c.)} = -0.45$, mid-band $A_V = -260$, $f_o = 2.79\,\text{kHz}$, $f_1 = 4.8\,\text{Hz}$

4.4 (a) $403\,\text{k}\Omega$ (b) $4.94\,\text{k}\Omega$ (c) 0.9975, $5\,\Omega$

4.5 With $R_S = 1\,\text{k}\Omega \| 5\,\text{k}\Omega$, $A_V = 0.83$, $r_{out} = 9.11\,\Omega$

4.6 (a) $A_V = -5.63$ (b) $V_{out} = 7.7\,\text{V}$ (c) $r_{in} = 10.8\,\text{k}\Omega$, $r_{out} = 36.75\,\Omega$
low frequency response:
$f_1 \approx 1.5\,\text{Hz}$ (owing to C_1)
$\left.\begin{matrix} f_2 \approx 44\,\text{Hz} \\ f_3 \approx 16\,\text{Hz} \end{matrix}\right\}$ (owing to C_2)

4.7 Analysis

4.9 $\beta = \beta_1 + \beta_2(1 + \beta_1)$, $r_{be} = r_{be1} + (1 + \beta_1)r_{be2}$

4.10 $A_V = -2.24$, $r_{in} = 54.6\,\text{k}\Omega$ ($I_C \approx 2\,\text{mA}$ and $R_E' = 670\,\Omega$)

5.1

V_1 (V)	V_2 (V)	V_3 (V)	TR1	TR2
1.5	1.3	5.0	off	active
2.0	1.3	4.25	active	active
2.5	1.8	3.5	active	off
3.0	2.3	3.5	active	off

5.2

V_1 (V)	V_2 (V)	V_3 (V)	TR1	TR2
1.5	1.3	5.0	off	active
2.0	1.3	4.025	active	active
2.5	1.8	2.3	active	off
3.0	2.3	≈ 2.3	saturated	off

5.3 CMRR = 186, input (a)

5.4 Tail current source resistance = $500\,\text{k}\Omega$. This is given by $R_E \geqslant 33.2\,\Omega$.

5.5 (a) input b, (b) CMRR = 1.8×10^3

5.6 $50\,\Omega$ ($25\,\Omega$ each)
v_A is amplitude $0.5 \times v_s$ centred on quiescent level 1.3 V
v_B is amplitude v_s (inverted) centred on quiescent level 3.6 V

5.7 Current mirror ratio = 0.806 or 1.025. The mismatch may be either way round and do not forget the $\beta/(\beta + 2)$ factor.

5.8 $R_2 = 15\,\text{k}\Omega$

5.9 $R_1 = 1\,\text{k}\Omega$ and $R_2 = 2.2\,\text{k}\Omega$

5.10 UTL = 614 mV, LTL = -88 mV, hysteresis = UTL $-$ LTL = 702 mV

6.1 $R_C = 4.7\,\text{k}\Omega$, $R_B = 47\,\text{k}\Omega$ (overdrive factor ≈ 2)

6.2 Transistors fully driven with overdrive of $2\times$. Therefore, two-state switch. OR function; either $V_{in(1)} = +10\,V$ or $V_{in(2)} = +10\,V$ or both turns ON TR1 and TR2 OFF with $V_{out} = +10\,V$.

6.3 V_{C3} is a positive pulse of amplitude 2.5 V and duration 1.69 ms, starting 0.77 ms after the positive-going edge of V_{in}

6.4 Output pulse duration $= 362\,\mu s$

6.5 $V_{CC} = +4.5\,V$, $R_C = 2.2\,k\Omega$, $R_T = R_{B1} = R_{B2} = 47\,k\Omega$, $C = 27\,nF$

6.6 Frequency $= 2.28\,kHz$, positive pulse duration $= 77\,\mu s$

6.7 Frequency rises with temperature increase

6.8 $V_{CC} = +4\,V$, $R_C = 1\,k\Omega$, $R_{B1} = R_{B2} = 27\,k\Omega$, $C1 = C_2 = 4.7\,nF$

6.9 Breakdown protection necessary, emitter diodes satisfactory. $V_{CC} = +10.7\,V$, $R_C = 1.8\,k\Omega$, $R_B = 56\,k\Omega$, $C_1 = 3.9\,nF$, $C_2 = 1\,nF$. There is a problem of selecting preferred value capacitors of 4:1 ratio along with a preferred value of R_B to give the correct frequency.

6.10 $R_A = 15\,k\Omega$, $R_B = 10\,k\Omega$, $C = 33\,nF$. The resistors can be scaled by a factor provided that the capacitor is also scaled by the reciprocal of that factor.

6.11 $k \approx 0.25$ so, using $R_1 = 1\,k\Omega$ and $R_2 = 3.3\,k\Omega$, $k = 0.233$. $t_1 = t_2 = 0.474 \times CR = 50\,\mu s$. Use $R = 10\,k\Omega$ and $C = 10\,nF$. With $V_{OH} = -V_{OL}$, the mark-to-space ratio is unity.

6.12 $t_1 = CR_B \ln \left[\dfrac{2V_{CC} - 2.1}{V_{CC} - 1.4} \right]$

6.13 $t_1 = CR_B \ln \left[\dfrac{2V_{CC} - 1.4}{V_{CC} - 1.4} \right]$

7.1 Intersect at $V_{GS} = 0.5V_p$. Pinch-off defined as: $I_{DS} = 0$ when $V_{GS} = V_p$. Cannot use this to measure V_p since I_{DS} is never zero owing to leakage. Using the above result, V_p can be calculated from accurately measured values of I_{DSS} and g_{mo}.

7.2 $w = 5.65\,\mu m$

7.3 (a) I_{DS} range is 1 mA to 6 mA (b) $V_{GS} = -0.25\,V$, I_{DS} is 1.5 to 7 mA.

7.4 $V_{GG} = +0.65\,V$, $R_S = 220\,\Omega$, R_G cannot be determined unless an input resistance specification is given

7.5 $r_{out} = 40\,k\Omega$

7.6 $A_V = 0.91$, $r_{out} = 298\,\Omega$

7.7 Differential gain $= 3.85$, CMG $= 0.0275$, CMRR $= 140$

7.9 $V_{DD} = +8\,V$, $\beta_L = 8\,\mu S/V$, $\beta_D = 28.6\,\mu S/V$

7.10 $V_{DD} = +8\,V$, $\beta_L = 8\,\mu S/V$, $\beta_D = 85.8\,\mu S/V$

8.1 (a) $I_{pk} = 0.75\,A$ (b) $P_{L(max)} = 2.25\,W$

8.2 $V_{CC} = 57\,V$

8.3 With bootstrapping, $A_V = -25\,233$
Without bootstrapping, $A_V = -789$

8.4 $V_{CC} = 28.5\,V$, $I_{pk} = 3.54\,A$

9.1 regulation $= 8.3\%$

9.2 $C = 666\,\mu F$ (use $1000\,\mu F$ to accommodate tolerance)

9.3 C doubled to $2200\,\mu F$ (often electrolytics available in only E3 series)

9.4 $L_{crit} = 0.255$ H, use 0.5 H. C then $6.76\,\mu$F, use $10\,\mu$F

9.5 D is 5.1 V, 400 mW. $R = 330\,\Omega$, 250 mW

9.6 ripple voltage = 83.8 mV peak-to-peak, $r_0 = 27.5\,\Omega$

9.7 6.2 V, 400 mW diode and 1 W BJT. $R_1 = 39\,\Omega$, 1.5 W. $R_2 = 68\,\Omega$, 250 mW.

9.8 $R = 680\,\Omega$, 250 mW. 4.7 V, 400 mW diode. $P_C = 600$ mW (absolute min)

9.9 $r_0 = 1.06\,\Omega$, RRF = 0.042

B.1

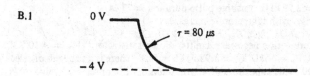

B.2

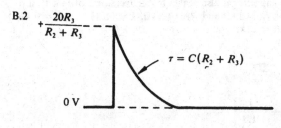

Ignore R_1 since V_{in} is a voltage source

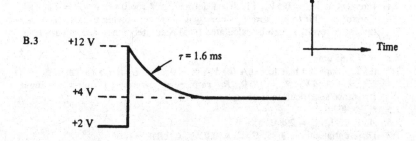

B.3

B.4

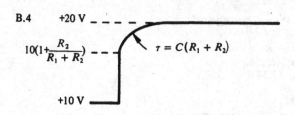

Index